Nanostructured Materials for Solar Cell Applications

Nanostructured Materials for Solar Cell Applications

Editor

Katsuaki Tanabe

MDPI • Basel • Beijing • Wuhan • Barcelona • Belgrade • Manchester • Tokyo • Cluj • Tianjin

Editor
Katsuaki Tanabe
Department of Chemical
Engineering
Kyoto University
Kyoto
Japan

Editorial Office
MDPI
St. Alban-Anlage 66
4052 Basel, Switzerland

This is a reprint of articles from the Special Issue published online in the open access journal *Nanomaterials* (ISSN 2079-4991) (available at: www.mdpi.com/journal/nanomaterials/special_issues/solar_cell_app).

For citation purposes, cite each article independently as indicated on the article page online and as indicated below:

LastName, A.A.; LastName, B.B.; LastName, C.C. Article Title. *Journal Name* **Year**, *Volume Number*, Page Range.

ISBN 978-3-0365-2865-6 (Hbk)
ISBN 978-3-0365-2864-9 (PDF)

Contents

Preface to "Nanostructured Materials for Solar Cell Applications"

The use of nanomaterials in technologies for photovoltaic applications continues to represent an important area of research. There are numerous mechanisms by which the incorporation of nanomaterials can improve device performance. We invited authors to contribute articles covering the most recent progress and new developments in the design and utilization of nanomaterials for highly efficient, novel devices relevant to solar cell applications. This book covers a broad range of subjects, from nanomaterials synthesis to the design and characterization of photovoltaic devices and technologies with nanomaterial integration.

Katsuaki Tanabe
Editor

Editorial

Nanostructured Materials for Solar Cell Applications

Katsuaki Tanabe

Department of Chemical Engineering, Kyoto University, Nishikyo, Kyoto 615-8510, Japan; tanabe@cheme.kyoto-u.ac.jp

The use of nanomaterials in technologies for photovoltaic applications continues to represent an important area of research. There are numerous mechanisms by which the incorporation of nanomaterials can improve device performance. We invited authors to contribute original research articles or comprehensive review articles covering the most recent progress and new developments in the design and utilization of nanomaterials for highly efficient, novel devices relevant to solar cell applications. This Special Issue aimed to cover a broad range of subjects, from nanomaterials synthesis to the design and characterization of photovoltaic devices and technologies with nanomaterial integration.

Yanbin Wang, Changlong Zhuang, Yawen Fang, Hyung Do Kim, Huang Yu, Biaobing Wang and Hideo Ohkita of Changzhou University, China and Kyoto University, Japan presented "Improvement of Exciton Collection and Light-Harvesting Range in Ternary Blend Polymer Solar Cells Based on Two Non-Fullerene Acceptors" [1]. Alvien Ghifari, Dang Xuan Long, Seonhyoung Kim, Brian Ma and Jongin Hong of Chung-Ang University, Korea presented "Transparent Platinum Counter Electrode Prepared by Polyol Reduction for Bifacial, Dye-Sensitized Solar Cells" [2]. Tianyi Shen, Qiwen Tan, Zhenghong Dai, Nitin P. Padture and Domenico Pacifici of Brown University, USA presented "Arrays of Plasmonic Nanostructures for Absorption Enhancement in Perovskite Thin Films" [3]. Qiang Zhang, Shengwen Hou and Chaoyang Li of Kochi University of Technology, Japan presented "Titanium Dioxide-Coated Zinc Oxide Nanorods as an Efficient Photoelectrode in Dye-Sensitized Solar Cells" [4]. Yasushi Shoji, Ryo Tamaki and Yoshitaka Okada of National Institute of Advanced Industrial Science and Technology, Japan and The University of Tokyo, Japan presented "Temperature Dependence of Carrier Extraction Processes in GaSb/AlGaAs Quantum Nanostructure Intermediate-Band Solar Cells" [5]. Panus Sundarapura, Xiao-Mei Zhang, Ryoji Yogai, Kazuki Murakami, Alain Fave and Manabu Ihara of Tokyo Institute of Technology, Japan and Univ Lyon, France presented "Nanostructure of Porous Si and Anodic SiO_2 Surface Passivation for Improved Efficiency Porous Si Solar Cells" [6]. Mao-Qugn Wei, Yu-Sheng Lai, Po-Hsien Tseng, Mei-Yi Li, Cheng-Ming Huang and Fu-Hsiang Ko of National Chiao Tung University, Taiwan and Taiwan Semiconductor Research Institute, Taiwan presented "Concept for Efficient Light Harvesting in Perovskite Materials via Solar Harvester with Multi-Functional Folded Electrode" [7]. Kodai Kishibe, Soichiro Hirata, Ryoichi Inoue, Tatsushi Yamashita and Katsuaki Tanabe of Kyoto University, Japan presented "Wavelength-Conversion-Material-Mediated Semiconductor Wafer Bonding for Smart Optoelectronic Interconnects" [8]. The guest editor would like to thank all of the authors of this Special Issue for their contributions to its successful completion.

Funding: This research received no external funding.

Conflicts of Interest: The author declare no conflict of interest.

Citation: Tanabe, K. Nanostructured Materials for Solar Cell Applications. *Nanomaterials* **2022**, *12*, 26. https://doi.org/10.3390/nano12010026

Received: 12 December 2021
Accepted: 13 December 2021
Published: 23 December 2021

References

1. Wang, Y.; Zhuang, C.; Fang, Y.; Kim, H.D.; Yu, H.; Wang, B.; Ohkita, H. Improvement of Exciton Collection and Light-Harvesting Range in Ternary Blend Polymer Solar Cells Based on Two Non-Fullerene Acceptors. *Nanomaterials* **2020**, *10*, 241. [CrossRef] [PubMed]
2. Ghifari, A.; Long, D.X.; Kim, S.; Ma, B.; Hong, J. Transparent Platinum Counter Electrode Prepared by Polyol Reduction for Bifacial, Dye-Sensitized Solar Cells. *Nanomaterials* **2020**, *10*, 502. [CrossRef] [PubMed]
3. Shen, T.; Tan, Q.; Dai, Z.; Padture, N.P.; Pacifici, D. Arrays of Plasmonic Nanostructures for Absorption Enhancement in Perovskite Thin Films. *Nanomaterials* **2020**, *10*, 1342. [CrossRef] [PubMed]
4. Zhang, Q.; Hou, S.; Li, C. Titanium Dioxide-Coated Zinc Oxide Nanorods as an Efficient Photoelectrode in Dye-Sensitized Solar Cells. *Nanomaterials* **2020**, *10*, 1598. [CrossRef] [PubMed]
5. Shoji, Y.; Tamaki, R.; Okada, Y. Temperature Dependence of Carrier Extraction Processes in GaSb/AlGaAs Quantum Nanostructure Intermediate-Band Solar Cells. *Nanomaterials* **2021**, *11*, 344. [CrossRef] [PubMed]
6. Sundarapura, P.; Zhang, X.-M.; Yogai, R.; Murakami, K.; Fave, A.; Ihara, M. Nanostructure of Porous Si and Anodic SiO_2 Surface Passivation for Improved Efficiency Porous Si Solar Cells. *Nanomaterials* **2021**, *11*, 459. [CrossRef] [PubMed]
7. Wei, M.-Q.; Lai, Y.-S.; Tseng, P.-H.; Li, M.-Y.; Huang, C.-M.; Ko, F.-H. Concept for Efficient Light Harvesting in Perovskite Materials via Solar Harvester with Multi-Functional Folded Electrode. *Nanomaterials* **2021**, *11*, 3362. [CrossRef]
8. Kishibe, K.; Hirata, S.; Inoue, R.; Yamashita, T.; Tanabe, K. Wavelength-Conversion-Material-Mediated Semiconductor Wafer Bonding for Smart Optoelectronic Interconnects. *Nanomaterials* **2019**, *9*, 1742. [CrossRef] [PubMed]

Article

Improvement of Exciton Collection and Light-Harvesting Range in Ternary Blend Polymer Solar Cells Based on Two Non-Fullerene Acceptors

Yanbin Wang [1,*], Changlong Zhuang [1], Yawen Fang [1], Hyung Do Kim [2], Huang Yu [1], Biaobing Wang [1,*] and Hideo Ohkita [2,*]

1 Jiangsu Key Laboratory of Environmentally Friendly Polymeric Materials, School of Materials Science and Engineering, Jiangsu Collaborative Innovation Center of Photovolatic Science and Engineering, Changzhou University, Changzhou 213164, Jiangsu, China; 19085204172@smail.cczu.edu.cn (C.Z.); 18000472@smail.cczu.edu.cn (Y.F.); 18000198@smail.cczu.edu.cn (H.Y.)

2 Department of Polymer Chemistry, Graduate School of Engineering, Kyoto University, Katsura, Nishikyo, Kyoto 615-8510, Japan; hyungdokim@photo.polym.kyoto-u.ac.jp

* Correspondence: wangyanbin@cczu.edu.cn (Y.W.); biaobing@cczu.edu.cn (B.W.); ohkita@photo.polym.kyoto-u.ac.jp (H.O.)

Received: 25 December 2019; Accepted: 28 January 2020; Published: 29 January 2020

Abstract: A non-fullerene molecule named Y6 was incorporated into a binary blend of PBDB-T and IT-M to further enhance photon harvesting in the near-infrared (near-IR) region. Compared with PBDB-T/IT-M binary blend devices, PBDB-T/IT-M/Y6 ternary blend devices exhibited an improved short-circuit current density (J_{SC}) from 15.34 to 19.09 mA cm^{-2}. As a result, the power conversion efficiency (PCE) increased from 10.65% to 12.50%. With an increasing weight ratio of Y6, the external quantum efficiency (EQE) was enhanced at around 825 nm, which is ascribed to the absorption of Y6. At the same time, EQE was also enhanced at around 600–700 nm, which is ascribed to the absorption of IT-M, although the optical absorption intensity of IT-M decreased with increasing weight ratio of Y6. This is because of the efficient energy transfer from IT-M to Y6, which can collect the IT-M exciton lost in the PBDB-T/IT-M binary blend. Interestingly, the EQE spectra of PBDB-T/IT-M/Y6 ternary blend devices were not only increased but also red-shifted in the near-IR region with increasing weight ratio of Y6. This finding suggests that the absorption spectrum of Y6 is dependent on the weight ratio of Y6, which is probably due to different aggregation states depending on the weight ratio. This aggregate property of Y6 was also studied in terms of surface energy.

Keywords: exciton harvesting; ternary blend solar cells; non-fullerene; energy transfer; surface energy

1. Introduction

Polymer solar cells have been studied widely because of their excellent advantages, such as flexibility, being light weight, and involving simple large-scale fabrication [1–5]. Typically, the photoactive layer of polymer solar cells is composed of one donor and one acceptor. Currently, conjugated polymers are widely employed as a donor material, and fullerene derivatives or non-fullerene derivatives are employed as an acceptor material in most cases. Fullerene acceptors have been dominated in the past two decades, and offered the highest power conversion efficiencies (PCEs) [6–10]. In 2015, Zhan et al. designed a non-fullerene acceptor named IT-IC, and fabricated the device based on a low-bandgap polymer PCE10 and ITIC. As a result, they obtained a PCE of 6.80%, which was higher than that of the devices based on PCE10 and PCBM [11]. Since then, much more attention has been paid to non-fullerene acceptors [12–15], and a record PCE of 18% has been reported very recently [16]. Among the non-fullerene acceptors, ((2,2′-((2Z,2′Z)-((12,13-bis(2-ethylhexyl)-3,9-

diundecyl-12,13-dihydro-[1,2,5]thiadiazolo [3,4-*e*]thieno[2,"3'':4',5']thieno[2',3':4,5]pyrrolo[3,2-*g*]thieno [2',3':4,5]thieno[3,2-*b*] indole-2,10-diyl)bis(methanylylidene))bis(5,6-difluoro-3-oxo-2,3-dihydro-1*H*-indene-2,1-diyliden e))dimalononitrile),Y6 has been most widely studied recently, and hence a PCE of more than 16% has been reported by several groups [17–21]. The high PCE of devices based on Y6 is partly ascribed to its wide light-harvesting range of up to 1000 nm. However, there is a limitation in binary blend polymer solar cells even though non-fullerene acceptors exhibit a good light-harvesting property. This is because the optical absorption bandwidth of organic photoactive materials is typically limited to be about 200 nm at most. Therefore, it is still difficult even for polymer/non-fullerene binary solar cells to effectively harvest many more photons over a wide wavelength range from the visible to the near-IR region at the same time.

Ternary blend polymer solar cells could be a simple and effective way to expand the optical absorption range [22–34]. Consequently, improved photovoltaic performance has been reported for ternary blend polymer solar cells [24–34]. The photoactive layer of ternary blend polymer solar cells is composed of one donor and two acceptors or two donors and one acceptor. In most cases, it is desirable to keep the ratio of donor to acceptor in the ternary blend the same as that in the optimized binary blend cell because the hole and electron transport would be balanced. In other words, with increasing weight ratio of the third component, the photon harvesting of the original component should decrease [35–39]. For polymer/fullerene solar cells, such a trade-off relation has been overcome by efficient energy transfer, which improves the exciton harvesting in ternary blend solar cells [24,25,28,40]. For example, when a low-bandgap polymer, poly[(4,4-bis(2-ethylhexyl)-dithieno[3,2-*b*:2',3'-*d*]silole)-2,6-diyl-*alt*-(2,1,3-benzothiadiazole)-4,7-diyl] (PSBTBT), was incorporated into binary blend devices based on poly(3-hexylthiophene) (P3HT) and a fullerene derivative (PCBM), the photocurrent was improved not only at the PSBTBT absorption band in the near-IR region but also at the P3HT absorption band in the visible region, although the optical absorption of P3HT rather decreased. This is because P3HT excitons lost in P3HT/PCBM binary blends can be collected by an efficient energy transfer from P3HT to PSBTBT, followed by an efficient charge transfer to PCBM [28].

In this study, a low-bandgap non-fullerene molecule named Y6 was incorporated into a binary blend of poly[[4,8-bis[5-(2-ethylhexyl)-2-thienyl]benzo[1,2-*b*:4,5-*b'*]dithiophene-2,6-diyl]-2,5-thiophenediyl [5,7-bis(2-ethylhex-yl)-4,8-dioxo-4*H*,8*H*-benzo[1,2-*c*:4,5-*c'*]dithiophene-1,3-diyl]] (PBDB-T) and 3,9-bis (2-methylene-(3-(1,1-dicyanomethylene)-6/7-methyl)-indanone))-5,5,11,11-tetrakis(4-*n*-hexylphenyl)-dithieno[2,3-*d*:2',3'-*d'*]-*s*-indaceno[1,2-*b*:5,6-*b'*]dithiophene (IT-M) to further improve the photon harvesting efficiency in the near-IR range. Figure 1 shows the chemical structures and the energy-level diagram of the three materials employed in this study. The complementary absorption also gives a large spectral overlap between the IT-M fluorescence and the Y6 absorption, which could enhance the exciton harvesting of IT-M by an efficient energy transfer from IT-M to Y6. Based on this strategy, a high short-circuit current density (J_{SC}) and a high fill factor (FF) were obtained at the same time, resulting in an improved PCE of 12.5%, which is even higher than those of both individual binary solar cells based on PBDB-T/IT-M and PBDB-T/Y6. We also found that PBDB-T/IT-M/Y6 ternary blends exhibit increased and red-shifted absorption in the near-IR region with increasing weight ratio of Y6. In order to address the origin of this spectral change, the absorption spectra of Y6 in different polymer matrices were studied in terms of surface energy.

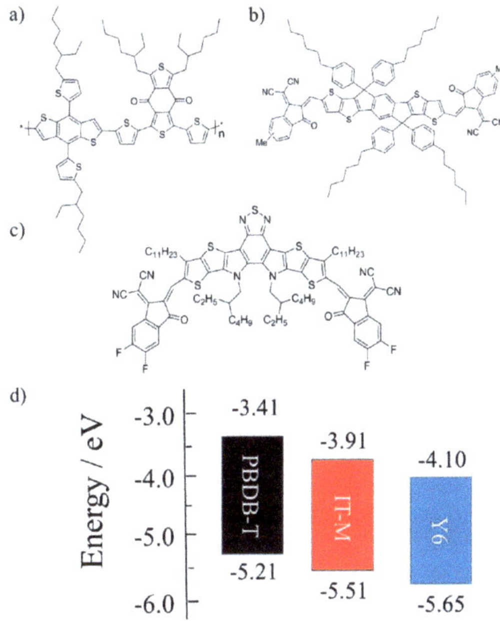

Figure 1. Chemical structures of materials employed in this study: (**a**) PBDB-T, (**b**) IT-M, and (**c**) Y6. (**d**) Energy level diagram of photovoltaic materials used in this study [11,18].

2. Materials and Methods

2.1. Materials

A conjugated polymer PBDB-T (Number average molecular weight M_n = 29 000 g mol^{-1}, polydispersity index PDI = 1.5), non-fullerene molecules IT-M (Purity: 99%) and Y6 (Purity: 99%), and poly[9,9-bis(3′-(*N*,*N*-dimethyl)-*N*-ethylammonium-propyl-2,7-fluorene)-*alt*-2,7-(9,9-dioctyl fluorine)] dibromide (PFN-Br) (M_n = 50,000 g mol^{-1}, PDI = 2.5) were purchased from Solarmer Materials, Incorporated (Beijing, China). These materials were used without further purification.

2.2. Device Fabrication

Non-fullerene-based ternary blend polymer solar cells were fabricated as follows. Indium–tin-oxide (ITO)-coated glass substrates (10 Ω per square) were rinsed by ultrasonication in toluene, acetone, and ethanol for 15 min in sequence. The cleaned substrates were dried under nitrogen gas and then treated with a UV–O_3 cleaner for 30 min. A hole-transporting buffer layer (40 nm) of PEDOT:PSS (AI4083) was spin coated onto the cleaned substrates at a spin rate of 3000 rpm for 60 s and then dried on a hot plate at 140 °C for 10 min in air. Prior to the spin coating, the solution of PEDOT:PSS was filtered with a PTFE syringe filter (pore size: 0.45 µm). The blend active layer was prepared on the ITO/PEDOT:PSS-coated substrate by spin coating at a spin rate of 2200 rpm for 60 s, and subsequently annealed on a hot plate at 140 °C for 10 min in the nitrogen atmosphere. A blend solution of PBDB-T/IT-M/Y6 was prepared by dissolving donor polymers (PBDB-T) and acceptors (IT-M and Y6) (1 : 1 by weight) with a composition of 10 : x : 10 − x mg in 1 mL of chlorobenzene with 1% volume ratio of 1,8-diiodooctane (DIO). The blend solution was stirred at 60 °C overnight. Note that the weight fraction of Y6 was optimized in the range 5–50 wt %. The thickness of blend films was ~100 nm. A PFN-Br buffer layer (~5 nm) was prepared on the active layer by spin coating at a spin rate of 3000 rpm for 60 s from a solution of PFN-Br (0.5 mg) in 1 mL anhydrous methanol. Finally, 100 nm of aluminum top electrode was thermally evaporated on top of the PFN-Br layer under vacuum at 2.5×10^{-4} Pa. Consequently, the device layered structure obtained was as follows: ITO/PEDOT:PSS/PBDB-T : IT-M : Y6/PFN-Br/Al. The effective area of the device was 0.07 cm^2.

2.3. Measurements

J–V characteristics were measured with a direct-current (DC) voltage and current source/monitor (Keithley, 2611B, Cleveland, USA) in the dark and under illumination with an AM 1.5G simulated solar light with 100 mW cm^{-2}. The light intensity was corrected with a calibrated silicon photodiode reference cell (Bunkou Keiki, BS-520, Tokyo, Japan). External quantum efficiency (EQE) spectra were measured with a spectral response measurement system (Bunkou Keiki, ECT-250D). The power of the incident monochromatic light was kept under 0.05 mW cm^{-2}, which was measured with a calibrated silicon reference cell (Bunkou Keiki, BS-520BK, Tokyo, Japan).

The ionization potential of PBDB-T, IT-M, and Y6 films was measured with a photoelectron yield spectrometer (Riken Keiki, AC-3, Tokyo, Japan). All the neat films were fabricated by spin coating from each chlorobenzene solution on the ITO substrate. The threshold energy for the photoelectron emission was estimated on the basis of the cubic root of the photoelectron yield plotted against the incident photon energy, as reported previously [24].

Absorption and photoluminescence (PL) spectra were measured at room temperature with a spectrophotometer (Hitachi, U-4100, Tokyo, Japan) and a spectrofluorometer (Horiba Jobin Yvon, NanoLog, Kyoto, Japna) equipped with a photomultiplier tube (Hamamatsu Photonics, R928P, Hamamatsu, Japan) and a liquid-nitrogen-cooled InGaAs near-IR array detector (Horiba Jobin Yvon, Symphony II, Kyoto, Japan), respectively.

The surface energy γ_X of the material X was evaluated from a contact angle θ_X, as reported previously [27,32,41]. The contact angle θ_X was measured for an ultrapure water droplet on the material film at room temperature. The interfacial energy γ_{AB} between materials A and B was evaluated from γ_A and γ_B by the Neumann's Equation.

3. Results

3.1. Optoelectronic Properties

As shown in Figure 2a, the donor polymer PBDB-T exhibits absorption bands in the visible region from 450 to 710 nm, the non-fullerene acceptor IT-M exhibits an absorption band in the visible to near-IR region from 500 to 800 nm, and the non-fullerene acceptor Y6 exhibits absorption bands mainly in the near-IR region with an absorption tail extending up to 1000 nm. In other words, these three photovoltaic materials show a complementary absorption from the visible to the near-IR range. On the other hand, as shown in Figure 2b, PBDB-T exhibits a PL peak at around 690 nm, IT-M exhibits a PL peak at around 770 nm and a shoulder peak at around 830 nm, and Y6 exhibits a PL peak at around 940 nm. Obviously, the PL peaks of PBDB-T and IT-M were located in the absorption range of IT-M and Y6, respectively. In other words, there is a good spectral overlap between the PL of PBDB-T and the absorption of IT-M, and between the PL of IT-M and the absorption of Y6, respectively, suggesting that the energy transfer from PBDB-T to IT-M and from IT-M to Y6 could occur. On the other hand, as shown in Figure 1d, cascade energy structures would be formed in PBDB-T/IT-M/Y6 blend films, which are beneficial for the charge transfer among the three materials. In other words, there is a competition of energy transfer with charge transfer at the interfaces of PBDB-T/IT-M and of IT-M/Y6.

Figure 3 shows the PL spectra of neat and blend films with different compositions upon photoexcitation of IT-M mainly at 710 nm. For the IT-M/Y6 binary blend film, the PL from IT-M was strongly quenched, and instead the PL from Y6 was clearly observed, indicating an efficient energy transfer from IT-M to Y6. For the PBDB-T/IT-M binary blend film, on the other hand, the PL from IT-M was quenched to ~10% relative to that of the IT-M neat film, suggesting that about 10% of IT-M excitons are radiatively deactivated to the ground state before arriving at the PBDB-T/IT-M interface. For the PBDB-T/IT-M/Y6 ternary blend film, no PL was observed by the addition of only 10 wt % of Y6 molecules into the PBDB-T/IT-M binary blends. In other words, the PL from IT-M was completely quenched and no PL from Y6 was observed. This is most probably because the 10% of IT-M excitons that would be lost in the absence of Y6 are collected to the Y6 domains by an energy transfer followed by a charge transfer to IT-M or PBDB-T.

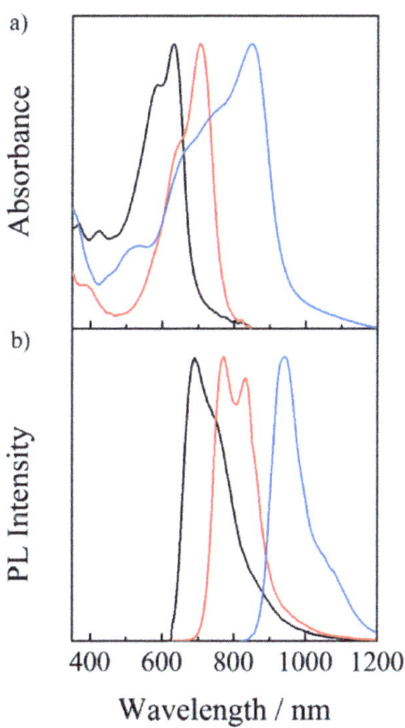

Figure 2. (**a**) Normalized UV-visible absorption and (**b**) PL spectra of PBDB-T (black lines), IT-M (red lines), and Y6 (blue lines) neat films excited at 550, 710, and 750 nm, respectively.

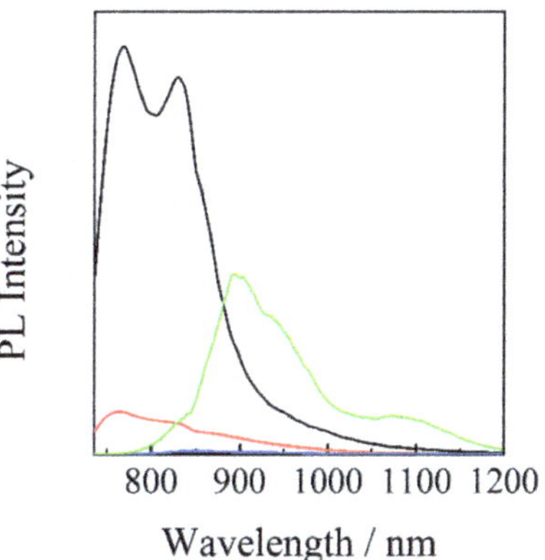

Figure 3. PL spectra of IT-M neat (black line), PBDB-T/IT-M (1 : 1) binary blend (red line), PBDB-T/IT-M/Y6 (1 : 0.8 : 0.2) ternary blend (blue line), and IT-M/Y6 (1 : 1) binary blend (green line) films. All the films were excited at 710 nm, and the PL intensity of all the films was corrected for variation in the absorption at an excitation wavelength of 710 nm. The PL intensity of Y6 in the IT-M/Y6 binary blend was corrected by subtracting the PL intensity due to the direct excitation of Y6 at 710 nm.

3.2. J–V Characteristics

In order to discuss the sensitization effect of Y6, we fabricated binary and ternary blend polymer solar cells with a structure of ITO/PEDOT:PSS/active layers/PFN-Br/Al under the same conditions. The device parameters are summarized in Table 1. The overall donor to acceptor ratio was maintained at 1 : 1 in this study. As shown in Figure 4, the PBDB-T/IT-M binary control device gave a short-circuit current density (J_{SC}) of 15.34 mA cm^{-2}, an open-circuit voltage (V_{OC}) of 0.946 V, a fill factor (FF) of 0.734, and a power conversion efficiency (PCE) of 10.65%, which are comparable to those reported previously [26,42]. With the incorporation of 10 wt % Y6 into the binary blend, the J_{SC} was obviously enhanced up to 19.09 mA cm^{-2}, which is much higher than that of the PBDB-T/IT-M binary control device and is approaching that of the PBDB-T/Y6 binary device. The FF was slightly decreased compared with that of the PBDB-T/IT-M binary control device, but it is much higher than that of the PBDB-T/Y6 binary device. V_{OC} was decreased compared with that of the PBDB-T/IT-M binary control device, but it is much higher than that of the PBDB-T/Y6 binary device. This is because the lowest unoccupied molecular orbital energy level of Y6 is much deeper than that of IT-M. As a result, the PCE was improved from 10.65% for the PBDB-T/IT-M binary control devices to 12.50% for the PBDB-T/IT-M/Y6 ternary blend devices, which is also much higher than that of the PBDB-T/Y6 binary blend devices. Further addition of Y6 rather decreased the photocurrent generation and hence degraded the overall photovoltaic performance.

Table 1. Photovoltaic parameters of PBDB-T/IT-M/Y6 ternary blend polymer solar cells with different compositions.

PBDB-T/IT-M/Y6	J_{SC}/mA cm^{-2}	V_{OC}/V	FF	PCE [a]/%
1 : 1 : 0	15.34 (15.02 ± 0.32)	0.946 (0.944 ± 0.002)	0.734 (0.718 ± 0.016)	10.65 (10.28 ± 0.27)
1 : 0.9 : 0.1	17.68 (17.24 ± 0.44)	0.917 (0.916 ± 0.001)	0.728 (0.701 ± 0.027)	11.72 (11.34 ± 0.38)
1 : 0.8 : 0.2	19.09 (18.72 ± 0.37)	0.902 (0.901 ± 0.001)	0.726 (0.713 ± 0.023)	12.50 (12.25 ± 0.22)
1 : 0.7 : 0.3	18.80 (18.37 ± 0.43)	0.872 (0.870 ± 0.001)	0.711 (0.693 ± 0.018)	11.65 (11.62 ± 0.33)
1 : 0 : 1	19.27 (19.05 ± 0.22)	0.689 (0.687 ± 0.002)	0.647 (0.625 ± 0.022)	8.60 (8.34 ± 0.26)

[a] The average values were obtained from 10 devices.

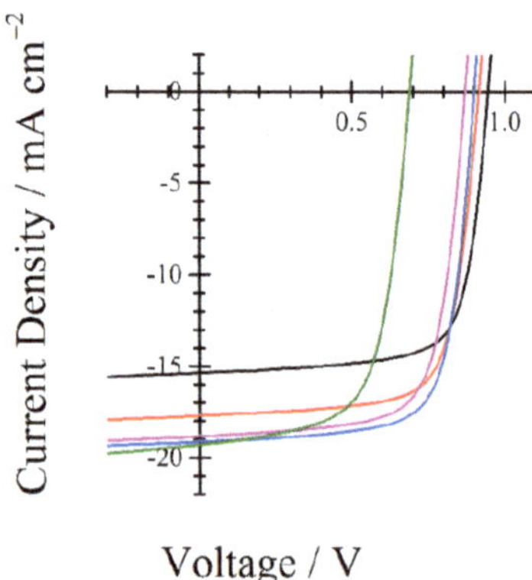

Figure 4. J–V characteristics of PBDB-T/IT-M/Y6 ternary blend polymer solar cells with different compositions: 1 : 1 : 0 (black line), 1 : 0.9 : 0.1 (red line), 1 : 0.8 : 0.2 (blue line), 1 : 0.7 : 0.3 (purple line), and 1 : 0 : 1 (green line).

3.3. External Quantum Efficiency (EQE) Spectra

In order to address the origin of enhancement in J_{SC}, we measured the absorption and EQE spectra of PBDB-T/IT-M/Y6 ternary blend solar cells with different compositions. As shown in Figure 5a, the absorption was increased at around 790 nm by the addition of Y6 into the binary blend of

PBDB-T/IT-M. Correspondingly, as shown in Figure 5b, EQE was enhanced near the Y6 absorption region. Instead, the absorption was decreased at around 700 nm, which is ascribed to the decrease of IT-M in the ternary blend. Interestingly, the EQE ascribed to IT-M was rather increased even though the absorption was decreased. More specifically, the EQE was increased at 700 nm from 72% to 82% by the incorporation of 10 wt % of Y6 into PBDB-T/IT-M binary blends, although the absorption efficiency was decreased at 700 nm from 92% to 87%. This is because there is efficient energy transfer from IT-M to Y6, as will be discussed detail later. Interestingly, as shown in Figure 5a,b, the absorption range was red-shifted with increasing weight ratio of Y6 in the ternary blend. Correspondingly, the EQE range was also red-shifted. This spectral change is probably due to the aggregation of Y6 in the ternary blend, which is dependent upon the weight fraction of Y6, as will be discussed later.

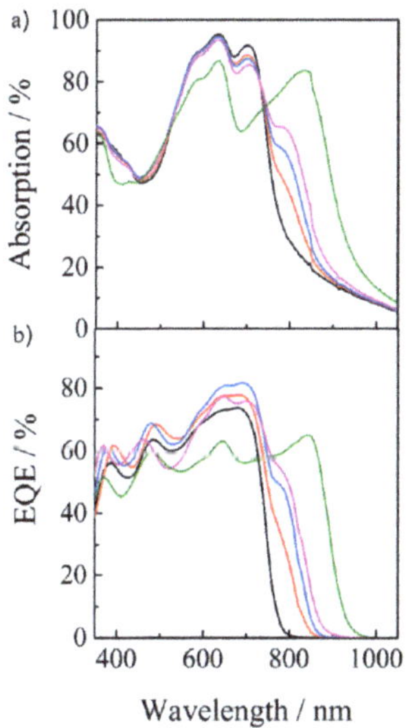

Figure 5. (**a**) UV-visible absorption and (**b**) EQE spectra of PBDB-T/IT-M/Y6 ternary blend films with different blend ratios: 1 : 1 : 0 (black line), 1 : 0.9 : 0.1 (red line), 1 : 0.8 : 0.2 (blue line), 1 : 0.7 : 0.3 (purple line), and 1 : 0 : 1 (green line).

In order to discuss the origin of such a spectral shift, we measured the absorption spectra of Y6 doped in different polymer films. As shown in Figure 6, the absorption spectra were different even though the Y6 weight fraction was the same (5 wt %), suggesting different aggregation states of Y6. In the chlorobenzene solution, Y6 isolated molecules exhibit an absorption band at around 730 nm. On the other hand, Y6 neat films exhibit absorption at around 843 nm, which is red-shifted by more than 100 nm compared to the absorption of Y6 in a chlorobenzene solution. In regiorandom poly(3-hexylthiophene) (RRa-P3HT) films, Y6 exhibits an absorption band at around 765 nm, which is similar to that of Y6 in a chlorobenzene solution, suggesting that Y6 are likely to be relatively homogeneously distributed in RRa-P3HT films. In polystyrene (PS) films, on the other hand, Y6 exhibits an absorption band at around 828 nm, which is rather similar to that in Y6 neat films, suggesting that Y6 molecules form aggregates similar to those in Y6 neat films. These findings suggest that Y6 molecules are more aggregated in these polymer films in the increasing order of PS, PBDB-T, and RRa-P3HT. These different aggregates will be discussed in terms of surface energy.

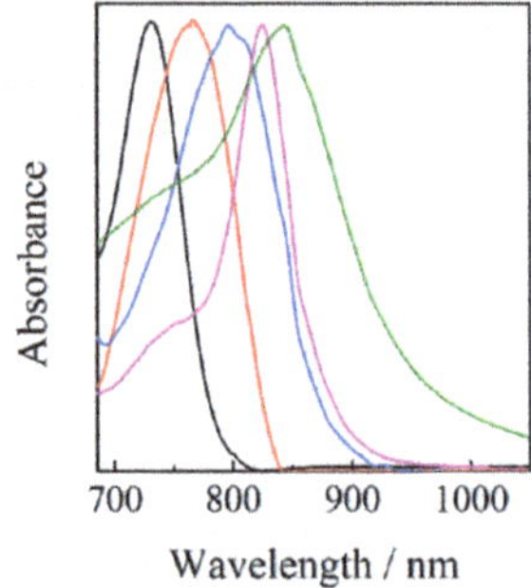

Figure 6. Normalized UV-visible absorption spectra of Y6 in a chlorobenzene solution (black line), Y6 neat film (green line), Y6 doped in RRa-P3HT (red line), PBDB-T (blue line), and polystyrene (PS) (purple line) films with a weight fraction of 5 wt %. The absorption spectra of Y6 in different polymer films were corrected by subtracting the absorption of polymer neat films.

4. Discussion

As mentioned above, PBDB-T/IT-M/Y6 ternary blend polymer solar cells exhibited the improved PCE compared to binary blend control cells. This is mainly due to the improved photocurrent by the addition of a near-IR non-fullerene molecule (Y6). Here, we discuss two mechanisms for the improved photocurrent: one is the improved exciton-harvesting of IT-M due to energy transfer and the other is the improved photon-harvesting by the addition of Y6.

Firstly, we focus on the energy transfer from IT-M to Y6. As shown in Figure 5, EQE was increased at around 700 nm while the absorption was rather decreased at around 700 nm, which is mainly ascribed to IT-M absorption. Thus, the internal quantum efficiency should be improved at the IT-M absorption. As shown in Figure 2, there is a good spectral overlap between the absorption of Y6 and the PL of IT-M, suggesting an efficient energy transfer from IT-M to Y6. Indeed, as shown in Figure 3, PL was observed from Y6 in the IT-M/Y6 binary blend even though IT-M was selectively excited, indicating an efficient energy transfer from IT-M to Y6. On the other hand, the PL of IT-M was quenched to 10% for the PBDB-T/IT-M binary blend, while it was completely quenched for the PBDB-T/IT-M/Y6 ternary blend, suggesting that the 10% IT-M excitons that would be lost in the PBDB-T/IT-M binary blend are efficiently collected to Y6 by energy transfer, as observed for the PL spectra of the IT-M/Y6 ternary blend. Subsequently, an efficient charge transfer occurs from Y6 to IT-M or PBDB-T. The improved exciton quenching efficiency reasonably explains the change of EQE and absorption.

Next, we focus on the improved photon harvesting by the addition of Y6. As shown in Figure 5, an additional EQE signal was observed at around 790 nm, which is consistent with an additional absorption due to the incorporation of Y6 into the PBDB-T/IT-M binary blend. This finding suggests that Y6 molecules contribute to the photocurrent generation in the ternary blend solar cell. Furthermore, the additional absorption was increased in intensity and also red-shifted in the absorption tail with increasing weight ratio of Y6. This is probable due to different aggregate states of Y6 depending on the weight fraction of Y6 in PBDB-T/IT-M/Y6 ternary blends.

In order to study this spectral property, the absorption spectra of Y6 were measured for different polymer matrices. As mentioned before, with the addition of 5 wt % of Y6 into RRa-P3HT, PBDB-T, and PS polymer matrices, the absorption peaks of Y6 were estimated to be 765 nm for Y6 in RRa-P3HT, 796 nm for Y6 in PBDB-T, and 828 nm for Y6 in PS films. This result suggests that the dispersed states in these three materials are different. With increasing weight ratio of Y6, as shown in Figure 7, the absorption peaks of Y6 in these three binary blends were red-shifted and finally saturated to a neat film state. In more detail, the peak shift was saturated at around 70 wt % for Y6 in RRa-P3HT, at around 50 wt % for Y6 in PBDB-T, and at around 20 wt % for Y6 in PS films. These results, again, indicate that Y6 molecules are likely to form aggregates more easily in the order of PS, PBDB-T, and RRa-P3HT films.

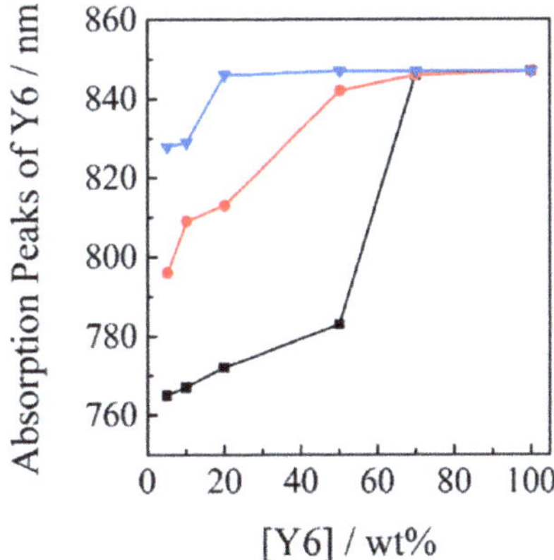

Figure 7. Peak wavelength of the Y6 absorption in different polymer matrices plotted against the weight fraction of Y6: RRa-P3HT (black line), PBDB-T (red line), and PS (blue line).

We therefore discuss the spectral property of Y6 molecules in terms of surface energy. Table 2 summarizes the surface energy of the materials used in this study. The compatibility of two different materials can be predicted on the basis of their surface energy: a similar surface energy indicates good compatibility, while a largely different surface energy indicates poor compatibility [43,44]. As shown in Table 2, the difference in surface energy $\Delta\gamma$ is as small as 2.6 mJ m^{-2} for Y6 and RRa-P3HT, which is the smallest, and the $\Delta\gamma$ is as large as 8.3 mJ m^{-2} for Y6 and PS, which is the largest. The trend in $\Delta\gamma$ is consistent with the absorption peak shift observed for Y6 in these polymer films. We thus conclude that dye aggregation can be controlled by the careful selection of material combinations.

Table 2. Surface energy (γ) of materials used in this study.

Materials	Y6	RRa-P3HT	PBDB-T	PS
γ/mJ m^{-2}	17.3	19.9	22.9	25.6

Finally, we consider the future direction for highly efficient non-fullerene polymer solar cells based on Y6. As shown in Figure 2, Y6 molecules exhibit a good optical absorption in the near-IR region up to 1000 nm, which is beneficial as a near-IR sensitizer for wide- and middle-bandgap photoactive materials. On the other hand, Y6 molecules also exhibit a large spectral change depending on the matrix material. In particular, the absorption of Y6 was most red-shifted in PS films with a large surface energy, resulting in the extension of the absorption tail up to 1000 nm. As shown in Figure 5a, the absorption peak of Y6 is 790 nm in PBDB-T/IT-M/Y6 ternary blend films with a Y6 fraction of 10 wt %. If we select a conjugated polymer with a higher surface energy like PS, Y6 molecules would exhibit a more red-shifted absorption band, resulting in a higher J_{SC}. We therefore believe that this finding is generally applicable to other photoactive materials with a large spectral change depending on the aggregation state.

5. Conclusions

In summary, PBDB-T/IT-M/Y6 ternary blend polymer solar cells exhibited J_{SC} increased from 15.34 to 19.09 mA cm^{-2}, and hence PCE improved from 10.65% to 12.5% compared to PBDB-T/IT-M binary blend polymer solar cells. The increased J_{SC} is partly due to the improved light harvesting by the additional absorption of Y6 in the near-IR region as well as the improved exciton harvesting by efficient energy transfer from IT-M to Y6. Interestingly, the absorption and EQE spectra due to Y6 were red-shifted with increasing weight ratio of Y6. We found that this special shift in absorption is dependent on the Y6 aggregation state, which is sensitive to the surface energy of component materials.

The absorption of Y6 was more red-shifted in polymer films with a surface energy more different from Y6, resulting in more efficient light harvesting up to 1000 nm. We thus believe that this finding is generally applicable to other photoactive materials with a large spectral change depending on the aggregation state.

Author Contributions: Y.W., B.W., H.D.K., and H.O. designed the research objective. Y.W. performed statistical analysis and drafted the initial article. C.Z., and Y.F. contributed to the section content and edited subsequent drafts, while all authors reviewed and provided feedstock on the submitted article. All authors have read and agreed to the published version of the manuscript.

Funding: This work was partly supported by the Natural Science Foundation of Jiangsu Province (BK20160280), Double Plan of Jiangsu Province (2016), Japan Science and Technology Agency (JST) ALCA program (JPMJAL1404), and Iketani Science and Technology Foundation (0314001-D).

Acknowledgments: We appreciate the Jiangsu Students' platform for innovation and entrepreneurship training program.

Conflicts of Interest: The authors declare no conflict of interest.

References

1. Nayak, P.K.; Mahesh, S.; Snaith, H.J.; Cahen, D. Photovoltaic solar cell technologies: Analysing the state of the art. *Nat. Rev. Mater.* **2019**, *4*, 269–285. [CrossRef]
2. Hou, W.; Xiao, Y.; Han, G.; Lin, J.Y. The applications of polymers in solar cells: A review. *Polymers* **2019**, *11*, 143. [CrossRef]
3. Li, Y.; Xu, G.; Cui, C.; Li, Y. Flexible and semitransparent organic solar cells. *Adv. Energy Mater.* **2018**, *8*, 1701791. [CrossRef]
4. Mateker, W.R.; McGehee, M.D. Progress in understanding degradation mechanisms and improving stability in organic photovoltaics. *Adv. Mater.* **2017**, *29*, 1603940. [CrossRef]
5. Dennler, G.; Scharber, M.C.; Brabec, C.J. Polymer-fullerene bulk-heterojunction solar cells. *Adv. Mater.* **2009**, *21*, 1323–1338. [CrossRef]
6. He, Z.; Xiao, B.; Liu, F.; Wu, H.; Yang, Y.; Xiao, S.; Wang, C.; Russell, T.P.; Cao, Y. Status and prospects for ternary organic photovoltaics. *Nat. Photonics* **2015**, *9*, 174–179. [CrossRef]
7. Liu, Y.; Zhao, J.; Li, Z.; Mu, C.; Ma, W.; Hu, H.; Jiang, K.; Lin, H.; Ade, H.; Yan, H. Aggregation and morphology control enables multiple cases of high-efficiency polymer solar cells. *Nat. Commun.* **2014**, *5*, 6293. [CrossRef]
8. Liang, Y.; Xu, Z.; Xia, J.; Tsai, S.T.; Wu, Y.; Li, G.; Ray, C.; Yu, L. For the bright future—bulk heterojunction polymer solar cells with power conversion efficiency of 7.4%. *Adv. Mater.* **2010**, *22*, E135–E138. [CrossRef]
9. Park, S.H.; Roy, A.; Beaupre, S.; Cho, S.; Coates, N.; Moon, J.S.; Moses, D.; Leclerc, M.; Lee, K.; Heeger, A.J. Bulk heterojunction solar cells with internal quantum efficiency approaching 100%. *Nat. Photonics* **2009**, *3*, 297–303. [CrossRef]
10. Li, G.; Shrotriya, V.; Huang, J.; Yao, Y.; Moriarty, T.; Emery, K.; Yang, Y. High-efficiency solution processable polymer photovoltaic cells by self-organization of polymer blends. *Nat. Mater.* **2005**, *4*, 864–868. [CrossRef]
11. Lin, Y.; Wang, J.; Zhang, Z.G.; Bai, H.; Li, Y.; Zhu, D.; Zhan, X. An electron acceptor challenging fullerenes for efficient polymer solar cells. *Adv. Mater.* **2015**, *27*, 1170–1174. [CrossRef]
12. Genene, Z.; Mammo, W.; Wang, E.; Andersson, M.R. Recent advances in n-type polymers for all-polymer solar cells. *Adv. Mater.* **2019**, *31*, 1807275. [CrossRef]
13. Wadsworth, A.; Moser, M.; Marks, A.; Little, M.S.; Gasparini, N.; Brabec, C.J.; Baran, D.; McCulloch, I. Critical review of the molecular design progress in non-fullerene electron acceptors towards commercially viable organic solar cells. *Chem. Soc. Rev.* **2019**, *48*, 1596–1625. [CrossRef]
14. Hou, J.; Inganäs, O.; Friend, R.H.; Gao, F. Organic solar cells based on non-fullerene acceptors. *Nat. Mater.* **2018**, *17*, 119–128. [CrossRef]
15. Chen, W.; Zhang, Q. Recent progress in non-fullerene small molecule acceptors in organic solar cells (OSCs). *J. Mater. Chem. C* **2017**, *5*, 1275–1302. [CrossRef]
16. Liu, Q.; Jiang, Y.; Jin, Y.; Qin, J.; Xu, J.; Li, W.; Xiong, J.; Liu, J.; Xiao, Z.; Sun, K.; et al. 18% efficiency organic solar cells. *Sci. Bull.* **2020**. [CrossRef]

17. Fan, B.; Zhang, D.; Li, M.; Zhong, W.; Zeng, Z.; Ying, L.; Huang, F.; Cao, Y. Achieving over 16% efficiency for single-junction organic solar cells. *Sci. China Chem.* **2019**, *62*, 746–752. [CrossRef]
18. Yuan, J.; Zhang, Y.; Zhou, L.; Zhang, G.; Yip, H.L.; Lau, T.K.; Lu, X.; Zhu, C.; Peng, H.; Johnson, P.A.; et al. Single-junction organic solar cell with over 15% efficiency using fused-ring acceptor with electron-deficient core. *Joule* **2019**, *3*, 1140–1151. [CrossRef]
19. Xu, X.; Feng, K.; Bi, Z.; Ma, W.; Zhang, G.; Peng, Q. Single-junction polymer solar cells with 16.35% efficiency enabled by a platinum (II) complexation strategy. *Adv. Mater.* **2019**, *31*, 1901872. [CrossRef]
20. Liu, B.; Wang, Y.; Chen, P.; Zhang, X.; Sun, H.; Tang, Y.; Liao, Q.; Huang, J.; Wang, H.; Meng, H.; et al. Boosting efficiency and stability of organic solar cells using ultralow-cost BiOCl nanoplates as hole transporting layers. *ACS Appl. Mater. Interfaces* **2019**, *11*, 33505–33514. [CrossRef]
21. Chang, Y.; Lau, T.K.; Pan, M.A.; Lu, X.; Yan, H.; Zhan, C. The synergy of host–guest nonfullerene acceptors enables 16%-efficiency polymer solar cells with increased open-circuit voltage and fill-factor. *Mater. Horiz.* **2019**, *6*, 2094–2102. [CrossRef]
22. Gasparini, N.; Salleo, A.; McCulloch, I.; Baran, D. The role of the third component in ternary organic solar cells. *Nat. Rev. Mater.* **2019**, *4*, 229–242. [CrossRef]
23. Rasi, D.D.C.; Janssen, R.A.J. Advances in solution-processed multijunction organic solar cells. *Adv. Mater.* **2019**, *31*, 1806499. [CrossRef] [PubMed]
24. Wang, Y.; Wang, T.; Chen, J.; Kim, H.D.; Gao, P.; Wang, B.; Iriguchi, R.; Ohkita, H. Quadrupolar D–A–D diketopyrrolopyrrole-based small molecule for ternary blend polymer solar cells. *Dyes Pig.* **2018**, *158*, 213–218. [CrossRef]
25. Wang, Y.; Chen, J.; Kim, H.D.; Wang, B.; Iriguchi, R.; Ohkita, H. Ternary blend solar cells based on a conjugated polymer with diketopyrrolopyrrole and carbazole units. *Front. Energy Res.* **2018**, *6*, 113. [CrossRef]
26. Wang, Y.; Kim, H.D.; Wang, B.; Ohkita, H. Visible sensitization for non-fullerene polymer solar cells using a wide bandgap polymer. *J. Photopolym. Sci. Technol.* **2018**, *31*, 177–181. [CrossRef]
27. Xu, H.; Ohkita, H.; Tamai, Y.; Benten, H.; Ito, S. Interface engineering for ternary blend polymer solar cells with a heterostructured near-IR dye. *Adv. Mater.* **2015**, *27*, 5868–5874. [CrossRef]
28. Wang, Y.; Ohkita, H.; Benten, H.; Ito, S. Highly efficient exciton harvesting and charge transport in ternary blend solar cells based on wide- and low-bandgap polymers. *Phys. Chem. Chem. Phys.* **2015**, *17*, 27217–27224. [CrossRef]
29. Wang, Y.; Zheng, B.; Tamai, Y.; Ohkita, H.; Benten, H.; Ito, S. Dye sensitization in the visible region for low-bandgap polymer solar cells. *J. Electrochem. Soc.* **2014**, *161*, D3093–D3096. [CrossRef]
30. Xu, H.; Ohkita, H.; Hirata, T.; Benten, H.; Ito, S. Near-IR dye sensitization of polymer blend solar cells. *Polymer* **2014**, *55*, 2856–2860. [CrossRef]
31. Xu, H.; Wada, T.; Ohkita, H.; Benten, H.; Ito, S. Dye sensitization of polymer/fullerene solar cells incorporating bulky phthalocyanines. *Electrochim. Acta* **2013**, *100*, 214–219. [CrossRef]
32. Honda, S.; Ohkita, H.; Benten, H.; Ito, S. Selective dye loading at the heterojunction in polymer/fullerene solar cells. *Adv. Energy Mater.* **2011**, *1*, 588–598. [CrossRef]
33. Honda, S.; Ohkita, H.; Benten, H.; Ito, S. Multi-colored dye sensitization of polymer/fullerene bulk heterojunction solar cells. *Chem. Commun.* **2010**, *46*, 6596–6598. [CrossRef] [PubMed]
34. Honda, S.; Nogami, T.; Ohkita, H.; Benten, H.; Ito, S. Improvement of the light-harvesting efficiency in polymer/fullerene bulk heterojunction solar cells by interfacial dye modification. *ACS Appl. Mater. Interfaces* **2009**, *4*, 804–810. [CrossRef]
35. Ameri, T.; Min, J.; Li, N.; Machui, F.; Baran, D.; Forster, M.; Schottler, K.J.; Dolfen, D.; Scherf, U.; Brabec, C.J. Performance enhancement of the P3HT/PCBM solar cells through NIR sensitization using a small-bandgap polymer. *Adv. Energy Mater.* **2012**, *2*, 1198–1202. [CrossRef]
36. Li, J.; Liang, Z.; Peng, Y.; Lv, J.; Ma, X.; Wang, Y.; Xia, Y. 36% Enhanced efficiency of ternary organic solar cells by doping a NT-based polymer as an electron-cascade donor. *Polymers* **2018**, *10*, 703. [CrossRef]
37. Zhang, M.; Zhang, F.; An, Q.; Sun, Q.; Wang, J.; Li, L.; Wang, W.; Zhang, J. High efficient ternary polymer solar cells based on absorption complementary materials as electron donor. *Sol. Energy Mater. Sol. Cells* **2015**, *141*, 154–161. [CrossRef]
38. Shen, W.; Chen, W.; Zhu, D.; Zhang, J.; Xu, X.; Jiang, H.; Wang, T.; Wang, E.; Yang, R. High-performance ternary polymer solar cells from a structurally similar polymer alloy. *J. Mater. Chem. A* **2017**, *5*, 12400–12406. [CrossRef]

39. An, Q.; Zhang, F.; Zhang, J.; Tang, W.; Wang, Z.; Li, L.; Xu, Z.; Teng, F.; Wang, Y. Enhanced performance of polymer solar cells through sensitization by a narrow band gap polymer. *Sol. Energy Mater. Sol. Cells* **2013**, *118*, 30–35. [CrossRef]
40. Wang, Y.; Ohkita, H.; Benten, H.; Ito, S. Efficient exciton harvesting through long-range energy transfer. *ChemPhysChem* **2015**, *16*, 1263–1267. [CrossRef]
41. Sumita, M.; Sakata, K.; Asai, S.; Miyasaka, K.; Nakagawa, H. Dispersion of fillers and the electrical conductivity of polymer blends filled with carbon black. *Polym. Bull.* **1991**, *25*, 265–271. [CrossRef]
42. Zhao, W.; Li, S.; Zhang, S.; Liu, X.; Hou, J. Ternary polymer solar cells based on two acceptors and one donor for achieving 12.2% efficiency. *Adv. Mater.* **2017**, *29*, 1604059. [CrossRef] [PubMed]
43. Huang, J.H.; Hsiao, Y.S.; Richard, E.; Chen, C.C.; Chen, P.; Li, G.; Chu, C.W.; Yang, Y. The investigation of donor-acceptor compatibility in bulk-heterojunction polymer systems. *Appl. Phys. Lett.* **2013**, *103*, 043304. [CrossRef]
44. Du, X.; Lin, H.; Chen, X.; Tao, S.; Zheng, C.; Zhang, X. Ternary organic solar cells with a phase-modulated surface distribution via the addition of a small molecular luminescent dye to obtain a high efficiency over 10.5%. *Nanoscale* **2018**, *10*, 16455–16467. [CrossRef] [PubMed]

Article

Transparent Platinum Counter Electrode Prepared by Polyol Reduction for Bifacial, Dye-Sensitized Solar Cells

Alvien Ghifari, Dang Xuan Long, Seonhyoung Kim, Brian Ma and Jongin Hong *

Department of Chemistry, Chung-Ang University, 84 Heukseok-ro, Dongjak-gu, Seoul 06974, Korea; alvien.ghifari@gmail.com (A.G.); dxlong.bk@gmail.com (D.X.L.); kimsh9560@gmail.com (S.K.); kehgeeng@gmail.com (B.M.)

* Correspondence: hongj@cau.ac.kr; Tel.: +82-2-820-5869

Received: 7 February 2020; Accepted: 3 March 2020; Published: 11 March 2020

Abstract: Pt catalytic nanoparticles on F-doped SnO_2/glass substrates were prepared by polyol reduction below 200 °C. The polyol reduction resulted in better transparency of the counter electrode and high power-conversion efficiency (PCE) of the resultant dye-sensitized solar cells (DSSCs) compared to conventional thermal reduction. The PCEs of the DSSCs with 5 μm-thick TiO_2 photoanodes were 6.55% and 5.01% under front and back illumination conditions, respectively. The back/front efficiency ratio is very promising for efficient bifacial DSSCs.

Keywords: dye-sensitized solar cell; counter electrode; bifacial; platinum; ethylene glycol

1. Introduction

Dye-sensitized solar cells (DSSCs) have shown promise as low-cost photovoltaics compared to commercially available Si solar cells. They hold great potential for building-attached photovoltaics (BAPVs) and building-integrated photovoltaics (BIPVs), because of their adjustable color/transparency and superior performance in dim light [1–4]. Recently, bifacial DSSCs, which can convert incoming sunlight to electricity through both front and back sides, have been an attractive alternative for photovoltaic devices [5,6]. The standard DSSC consists of a substrate coated with transparent conducting oxides (TCOs), a dye-grafted mesoscopic TiO_2 photoanode, a platinized counter electrode (CE), and an electrolyte containing a redox couple. Among the key components, the CE plays a prominent role in maintaining a flow of current by regenerating the redox mediator. Unfortunately, Pt is a highly expensive and scarce metal, and thus alternative materials, including carbon-based materials [7,8], transition metal compounds [9], conducting polymers [10,11], and their composites [12,13] have been explored.

Nevertheless, Pt is still favored because of its superior electrocatalytic activity and good conductivity. The conventional methods for preparing Pt CEs are vacuum sputtering of a Pt target and thermal decomposition of a platinic acid (H_2PtCl_6) precursor [14,15]. However, these high-energy-consuming approaches increase the cost and energy payback time of DSSCs. In addition to logistical issues, the light absorption at the Pt CEs should be minimized for bifacial operation. Therefore, it is crucial to develop new preparation methods for highly transparent Pt CEs at low temperatures.

Polyol-based synthesis is a versatile and straightforward liquid-phase method that uses high-boiling and multivalent alcohols to synthesize nanomaterials without the requirements of high pressure and autoclaves [16,17]. Polyol can simultaneously act as a reducing agent and water-equivalent solvent. Its chelating ability or controlling the nucleation of nanomaterials. For example, Ouyang and coworkers reported nanostructured Pt CEs prepared by polyol reduction of H_2PtCl_6 in ethylene glycol

(EG) [18,19]. They also utilized EG vapor for solventless chemical reduction of the platinum precursor below 200 °C [20]. Song et al. employed the hydrolysis of urea in a two-step EG solution reduction to achieve uniform dispersion and a smaller size of Pt nanoparticles on conducting glass substrates [21]. Unfortunately, the Pt CEs were prepared by drop-casting the EG-based Pt precursor solution. It has proven challenging to prepare a thin layer of Pt nanoparticles and thus provide highly transparent CEs for the bifacial DSSCs. Therefore, we prepared a highly transparent Pt counter electrode by spin-coating a platinic EG solution and further chemical reduction below 200 °C. We also investigate the photovoltaic performance of the DSSCs from the viewpoint of bifacial operation.

2. Materials and Methods

2.1. Fabrication of Pt Counter Electrodes

F-doped SnO_2 (FTO) glass (TEC 8, Pilkington; sheet resistance = 8Ω/□) substrates were ultrasonically cleaned using acetone, isopropyl alcohol, and deionized water. The substrates were baked at 150 °C for 10 min to completely remove residual water. Then 10 mM platinic acid in ethylene glycol was spin-coated on the FTO substrates. Subsequently, the sample was placed in a muffle furnace and heated to a certain temperature (e.g., 130, 150, 170, 190, and 210 °C). The annealing was maintained for 12 h, and then the sample was cooled down to room temperature. For comparison, 40 mM of the platinic acid solution in 2-propanol was prepared and then spin-coated on the same substrates, followed by heating at 425 °C for 1 h in the muffle furnace [22–24]. The schematic diagram of Pt CE preparation is shown in Figure 1.

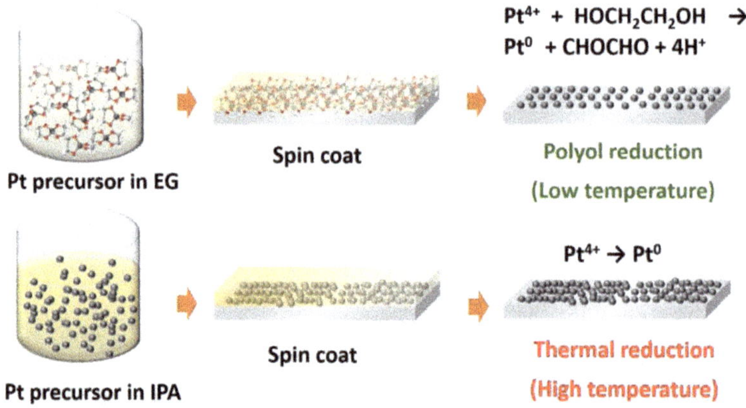

Figure 1. Schematic diagram of counter electrode preparation: polyol reduction (PR) and thermal decomposition (TD).

2.2. Characterization

The surface images of the fabricated Pt CEs were obtained using a field-emission scanning electron microscope (FE-SEM; Sigma, Carl Zeiss AG, Oberkochen, Germany). X-ray photoelectron spectroscopy (XPS) spectra were acquired using a Thermo Scientific K-alpha XPS system (Waltham, MA, United States) with an Al Kα X-ray source monochromator (1486.6 eV). The optical transmittance spectra were recorded using UV/Vis spectroscopy (V-730, JASCO, Tokyo, Japan). An electrochemical measurement station (CompactStat, Ivium Technologies, Eindhoven, The Netherlands) was used to conduct all electrochemical characterization. Cyclic voltammetry (CV) was recorded at a scan rate of 50 mV/s with a three-electrode system that consists of a Pt CE as a working electrode, a Pt wire as a counter electrode, and Ag/AgCl as a reference electrode. A solution of 10.0 mM LiI, 1.0 mM I_2, and 0.1 M $LiClO_4$ in CH_3CN was used as the electrolyte to investigate the electrocatalytic properties of the Pt CEs for redox reactions. Tafel and electrochemical impedance spectroscopy (EIS) measurements were performed on symmetric cells consisting of CE|electrolyte|CE. Tafel polarization measurements were conducted at a scan rate of 10 mV/s. In EIS measurements, a 10 mV amplitude sinusoidal potential perturbation was input over a frequency range from 1 MHz to 0.1 Hz at zero bias potential. The EIS spectra were analyzed using the equivalent circuit fitting routine in the ZView software (AMETEK, Leicester, UK).

2.3. Device Fabrication and Characterization

The FTO glass substrates were cleaned with O_2 plasma for 10 min, dipped in an aqueous solution of 40 mM $TiCl_4$ at 75 °C for 30 min, and then rinsed several times with deionized water. The TiO_2 paste (Transparent TiO_2, ENBKOREA, Gumi, Korea) was screen-printed on the FTO glass, and the printed film was calcinated at 300 °C for 30 min and 575 °C for 1 h in the muffle furnace. The final areas of the TiO_2 photoanodes were 0.2025 cm^2. The photoanodes were treated in the $TiCl_4$ solution and then heated at 500 °C for 30 min on a hot plate. After O_2 plasma treatment, the photoanodes were dipped into a 0.5 mM dye solution of cis-diisothiocyananoto-bis(2,2-bipyridyl-4,4′-dicarboxylate) ruthenium(II) (N719, Merck KGaA, Darmstadt, Germany) in ethanol for 24 h. The dye-grafted photoanode and Pt CE were assembled with 25 μm-thick thermoplastic film (Surlyn, Solaronix, Aubonne, Switzerland) and sealed by heating. An iodide-based redox electrolyte (Iodolyte AN-50, Aubonne, Switzerland) was injected into the pre-drilled holes in the side of the counter electrode and then sealed. The photovoltaic characteristics were investigated under AM 1.5 global one sun illumination (100 mW/cm^2) using a solar cell I–V measurement system (K3000 LAB, McScience, Suwon, Korea). The photocurrent density (J_{sc}), open-circuit voltage (V_{oc}), fill factor (FF), and power conversion efficiency (η) were recorded simultaneously. Monochromatic incident photon-to-current conversion efficiency (IPCE) was collected to evaluate the spectral response of the solar cells (K3100, McScience, Suwon, Korea). EIS measurement on the fabricated devices was carried out by the same protocol mentioned above.

3. Results

Polyol-based synthesis is a versatile technique to prepare Pt nanostructures. In this study, EG was chosen as the solvent and reducing agent of Pt precursors because of its low boiling point and viscosity. Also, byproducts from the EG reduction, such as glycoaldehyde and diacetyl, could be easily removed [25]. Figure 2a–f shows the surface of the FTO glass substrates decorated with Pt nanoparticles formed by polyol reduction at different temperatures (130, 150, 170, 190, and 210 °C) and thermal decomposition at 425 °C (hereafter, "polyol-reduced" is abbreviated as "PR" and "thermally decomposed" is abbreviated as "TD"). Tiny Pt nanoparticles (<10 nm) were formed and dispersed on the FTO surface for all reaction temperatures. The chemical reduction by EG allows for depositing Pt nanoparticles at low temperatures, and is thus feasible for plastic substrates. As the reduction temperature increased, the aggregation of the Pt nanoparticles diminished, and no aggregation could be observed at the temperature of 190 °C. This indicates that the slower evaporation of EG could result in the growth of larger nanoparticles, more agglomeration, and dendrites. However, the aggregated nanoparticles appeared again at a temperature higher than the boiling point of EG (i.e., 197 °C) because of the inability to control particle nucleation and growth at the elevated temperature [17]. Also, the pyrolysis of H_2PtCl_6 at 425 °C directly led to the formation of large Pt nanoparticles prominently populating the FTO. Accordingly, we predict that their size and distribution will affect the transparency and catalytic activity of the resultant electrodes.

XPS was performed to determine the formation of metallic Pt during our chemical reduction. In Figure 3, the spectra of survey XPS indicates that the samples contained Sn, O, Pt, and Cl—no other elements except carbon were detected. Carbon can result from precursors or sample handling. Interestingly, as the reduction temperature increased, the signal for Cl 2p related to ionic Pt species decreased and then disappeared. Narrow XPS scans of the Pt-4f core level region are also provided in Figure 3. The C 1 s peak at 285.0 eV was used to calibrate all the XPS data. The Pt 4f signal is composed of three pairs of deconvoluted doublets. The first doublet (71.4 eV and 74.8 eV) corresponds to the platinum in the zero-valent state (i.e., Pt(0)), while the second doublet (72.3 eV and 75.6 eV) can be assigned to platinum in the 2+ valence state (i.e., Pt(II)) [26]. The third (73.7 eV and 77.0 eV) is caused by the Pt^{4+} species, such as $[PtCl_6]^{2-}$, on the surface (i.e., Pt(IV)) [27]. Table 1 summarizes the binding energies and relative integrated peak areas of the deconvoluted peaks at the Pt-4f core level region. The increase in the reduction temperature resulted in the change in the valence state from Pt^{2+} to Pt^0. After the polyol reaction, the Pt species clustered together with different compositions. Accordingly, the reaction temperature is of significant importance in preparing metallic Pt nanoparticles.

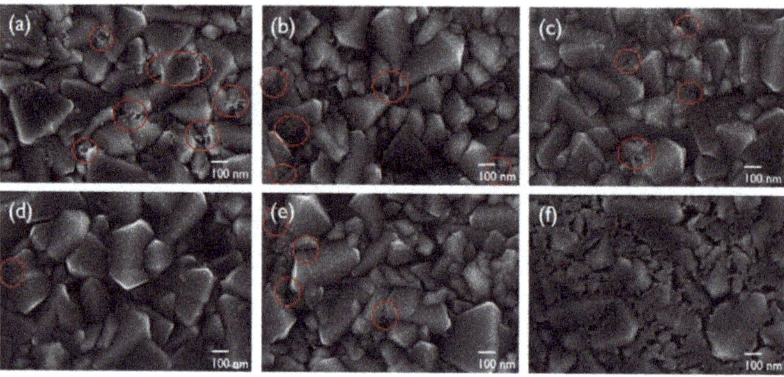

Figure 2. Field-emission scanning electron microscope (FE-SEM) images of Pt films on FTO substrates: (**a**) PR-130, (**b**) PR-150, (**c**) PR-170, (**d**) PR-190, (**e**) PR-210, and (**f**) TD-425.

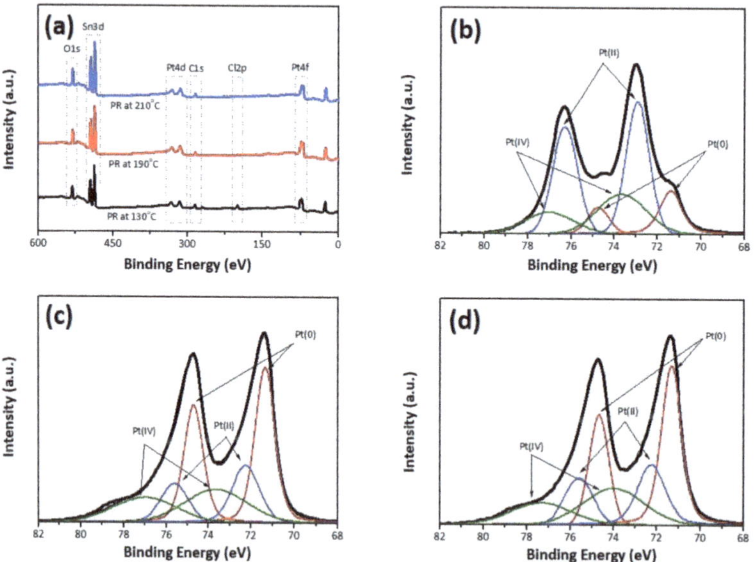

Figure 3. X-ray photoelectron spectroscopy (XPS) scans of polyol-reduced Pt on FTO substrates: (**a**) survey, (**b**) PR-130, (**c**) PR-190, and (**d**) PR-210.

Table 1. Binding energies (eV) and relative proportions (S, in %) of each component measured by XPS.

	Pt(0)			Pt(II)			Pt(IV)		
	$4f_{5/2}$	$4f_{7/2}$	S	$4f_{5/2}$	$4f_{7/2}$	S	$4f_{5/2}$	$4f_{7/2}$	S
PR-130	74.8	71.4	23.7	76.3	72.9	38.8	77.0	73.7	37.5
PR-190	74.7	71.4	48.4	75.6	72.3	22.7	77.0	73.7	28.9
PR-210	74.7	71.4	44.8	75.6	72.3	20.9	77.0	73.7	34.3

In the DSSC operation, an I_3^-/I^- redox couple is commonly utilized as a redox mediator. The CE should collect electrons from the external circuit and effectively catalyze the reduction of I_3^- to I^-. The electrocatalytic activity of the PR CEs was determined using cyclic voltammetry (CV), as shown in Figure 4a. All CV curves show two typical pairs of oxidation and reduction peaks (Ox-1/Re-1 and Ox-2/Re-2), which are described by Equations (1) and (2), respectively [28]:

$$I_3^- + 2e^- \rightleftharpoons 3I^- \tag{1}$$

$$3I_2 + 2e^- \rightleftharpoons 2I_3^- \tag{2}$$

The catalytic reduction activity of the CE can be determined by the first oxidation and reduction peaks (Ox-1 and Re-1). Its electrochemical reversibility can be assessed from the peak-to-peak separation (ΔE_p), which is the difference between the anodic and cathodic peak potentials. A smaller ΔE_p reflects the higher electrocatalytic activity of the CE.

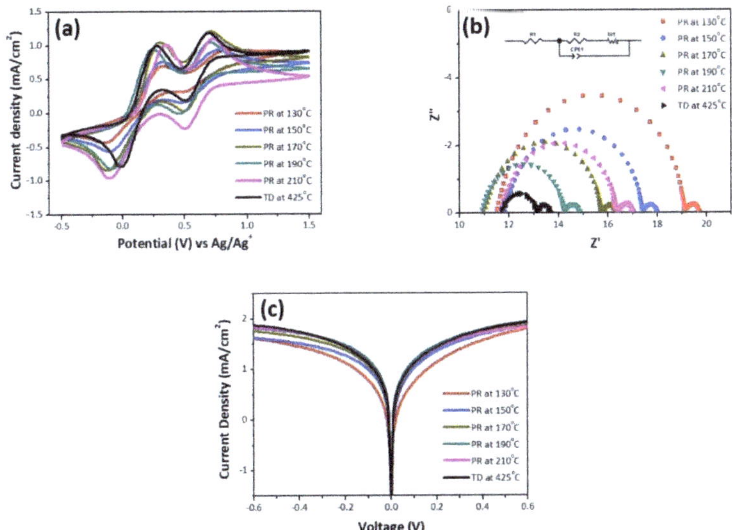

Figure 4. (**a**) Cyclic voltammograms for the redox of I_3^-/I^- species, (**b**) electrochemical impedance spectroscopy (EIS) Nyquist plots for the symmetric cells, and (**c**) Tafel polarization curves.

EIS measurements were also conducted using the symmetrical cells constructed with two identical electrodes. The Nyquist plots (Figure 4b) consist of two apparent semicircles at a high frequency (first semicircle) and a low frequency (second semicircle), respectively. The Randles-type circuit (insert in Figure 4b) was used to simulate the plots. The corresponding parameters are summarized in Table 2. R_s is the series resistance, and can be derived by the intercept of the high-frequency semicircle on the real axis (Z' axis). R_{ct} is the charge transfer resistance at the interface between CE and electrolyte, and can be determined from the radius of the first semicircle on the real axis. All the PR-CEs have nearly the same R_s, and thus its effect on photovoltaic performance can be ignored. A smaller R_{ct} accelerates triiodide reduction, and thus the PR-190 CE has the superior catalytic activity. The low-frequency semicircle results from Nernst diffusion impedance (Z_N) of the redox species in the electrolyte, which is inversely proportional to the diffusion coefficient of I_3^- in the cells [28]. The decrease in Z_N leads to an increase in the electrocatalytic activity of the CE.

Table 2. Electrochemical parameters of polyol-reduction Pt films at different reaction temperatures.

No	ΔE_p (V)	J_0 (mA/cm^2)	J_{lim} (mA/cm^2)	R_s (Ω/cm^2)	R_{ct} (Ω/cm^2)	Z_N (Ω/cm^2)
PR-130	0.47	4.99	0.21	8.05	5.19	0.44
PR-150	0.44	9.16	0.21	8.26	3.79	0.44
PR-170	0.43	11.14	0.24	7.69	3.21	0.41
PR-190	0.35	14.76	0.26	7.63	2.23	0.44
PR-210	0.46	12.44	0.26	8.13	3.17	0.48
TD-425	0.29	11.62	0.27	11.79	1.27	0.59

Figure 4c shows Tafel polarization plots of the symmetrical cells comprising the Pt CEs and I_3^-/I^- electrolytes. The Tafel plot can be separated into three consecutive zones: polarization, Tafel, and limit diffusion zones. The Tafel and limit diffusion zones are valuable for obtaining both the limiting current density (J_{lim}) and current density (J_0), which correlate with the electrocatalytic activity of the CEs [29]. The intersection of the cathodic branch and the equilibrium potential line can be considered J_0, and thus the steep Tafel slope implies a large J_0. Theoretically, J_0 can also be calculated using Equation (3):

$$J_0 = \frac{RT}{nFR_{ct}} \tag{3}$$

where R is the gas constant, T is the absolute temperature, n is the number of electrons participating in the electrochemical reduction of I_3^-, and F is Faraday's constant. In the limit diffusion zone, J_{lim} can be determined by the intersection of the cathodic branch and the y-axis. J_{lim} is directly proportional to a diffusion coefficient of I_3^- (D) at the same potential, and can be expressed as Equation (4):

$$J_{lim} = \frac{2neDCN_A}{l} \tag{4}$$

where e is the elementary charge, C is the concentration of I_3^-, N_A is the Avogadro constant, and l is the distance between two electrodes. Notably, the values of J_0 and J_{lim} followed the same trend observed in both CV and EIS analyses. Accordingly, the electrocatalytic activity of I_3^-/I^- is as follows: PR-190 > PR-170 > PR-210 > PR-150 > PR-130.

The light absorption at the counter electrode should be minimized to improve light harvesting in the bifacial DSSCs. Figure 5a shows the transmittance spectra of the prepared Pt CEs and bare FTO glass. Polyol reduction manifested higher transparency compared to thermal decomposition in the whole visible light regime, which will be beneficial to bifacial applications. Unfortunately, the increase in reduction temperature resulted in the decrease in transmittance of PR CEs. We think that metallic Pt nanoparticles possibly have a negative effect on the light transparency. Figure 5b,c shows, respectively, the current density-voltage (J–V) characteristics and IPCE spectra of the DSSCs (5 μm-thick TiO_2), which were illuminated from the TiO_2 photoanode side (i.e., front illumination). The photovoltaic parameters of the front- and rear-illuminated DSSCs are summarized in Table 3. The improved electrocatalytic activity of the PR–Pt CE resulted in better photovoltaic performance: power conversion efficiency (η) increased from 5.50% (PR-130) to 6.55% (PR-190). Also, PR-190 exhibited better photovoltaic performance than TD-425 because of the improvement of R_s and Z_N related to the electrocatalytic activity of the CE. It should be noted that all the DSSCs employing PR–Pt CEs maintained more than 76% of their front-illumination efficiency when lit from the Pt CE side (i.e., back illumination). The decrease of η in the back-illumination condition is related to the transmission losses due to the

Pt-based electrocatalyst and the I_3^-/I^- electrolyte [6]. The ratio of the back-illumination efficiency to the front-illumination efficiency ($\eta(R)$) followed the trend of the transmittance of the PR–Pt CEs observed above.

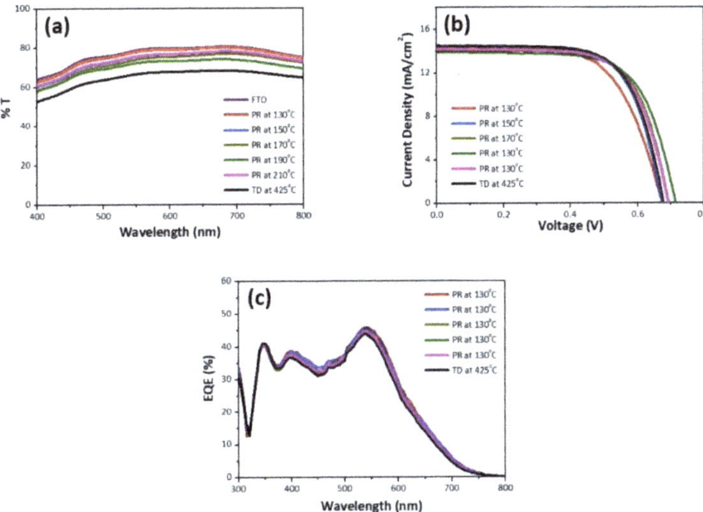

Figure 5. (**a**) Transmittance spectra of the Pt counter electrodes, (**b**) J–V characteristic curves of the dye-sensitized solar cells (DSSCs) (5 µm-thick TiO_2) under one sun front illumination, and (**c**) their incident photon-to-current conversion efficiency (IPCE) spectra.

Table 3. Photovoltaic parameters of DSSCs (5 µm-thick TiO_2) with different Pt counter electrodes [a].

No	Illumination	V_{oc} (V)	J_{sc} (mA/cm^2)	*FF* (%)	η (%)	η (*R*) (%)
PR-130	Front	0.666 ± 0.007	13.82 ± 0.23	59.70 ± 5.48	5.50 ± 0.66	81.03 ± 1.68
	Back	0.664 ± 0.008	10.67 ± 0.32	62.74 ± 4.85	4.45 ± 0.50	
PR-150	Front	0.673 ± 0.003	13.99 ± 0.51	67.45 ± 1.43	6.35 ± 0.33	77.57 ± 1.57
	Back	0.670 ± 0.004	10.65 ± 0.62	69.04 ± 1.22	4.93 ± 0.35	
PR-170	Front	0.694 ± 0.006	13.96 ± 0.31	67.20 ± 1.33	6.51 ± 0.30	76.73 ± 1.33
	Back	0.689 ± 0.003	10.62 ± 0.52	68.28 ± 0.90	5.00 ± 0.30	
PR-190	Front	0.702 ± 0.011	13.77 ± 0.12	67.66 ± 1.55	6.55 ± 0.28	76.61 ± 3.24
	Back	0.699 ± 0.00	10.44 ± 0.23	68.71 ± 1.76	5.01 ± 0.24	
PR-210	Front	0.695 ± 0.004	13.76 ± 0.45	67.02 ± 0.84	6.41 ± 0.29	76.70 ± 2.14
	Back	0.692 ± 0.008	10.42 ± 0.68	68.22 ± 0.94	4.92 ± 0.3	
TD-425	Front	0.682 ± 0.00	13.97 ± 0.46	67.36 ± 0.72	6.42 ± 0.28	67.92 ± 2.66
	Back	0.672 ± 0.005	9.40 ± 0.16	68.97 ± 0.66	4.35 ± 0.05	

[a] The parameters were obtained from at least five cells fabricated with each CE condition; each cell was measured five times.

To date, various transparent CEs have been developed for bifacial applications, such as photovoltaic windows and façades. Table 4 shows the photovoltaic performance of the bifacial DSSCs with platinum and non-platinum CE materials compared to our work. Although DSSC fabrication conditions (e.g., TiO_2 thickness, dyes, electrolytes) are not the same, our polyol reduction technique proved sufficient for fabricating bifacial DSSCs.

Table 4. Photovoltaic parameters of bifacial DSSCs with different CE materials.

Material	Method	T (%)	V_{oc} (V)	J_{sc} (mAcm^{-2})	*FF* (%)	*η (front)* (%)	*η (back)* (%)	*η (R)* (%)	Refs.
PEDOT	Electropolymerization	~90 [a]	0.751	14.60	67	7.40	5.23	70.7	[6]
Carbon	Carbonization	~75	0.721	10.52	60	6.07	5.04	83.0	[7]
Ni_3S_4	Hydrothermal	~70	0.700	13.58	65	6.56	4.86	73.9	[9]
PANI/MoS_2	Composite	~63	0.799	17.93	63	7.99	3.40	42.6	[12]
Pt	Photo-reduction	~95	0.810	13.53	66.6	7.29	5.85	80.3	[30]
Pt	Thermal decomposition	~80 [a]	0.757	15.00	70.7	8.02	4.43	55.2	[30]
Pt	Polyol reduction [b]	~80	0.702	13.77	67.6	6.55	5.01	76.6	-

[a] Transmittance was calculated from absorbance. [b] Our work.

4. Conclusions

In this work, highly transparent Pt CEs were prepared using the polyol reduction technique at low temperatures (<200 °C). Our facile and versatile technique provided better electrocatalytic activity and transparency than conventional thermal decomposition methods, and thus brought significant improvement to the photovoltaic performance of the bifacial DSSCs. In particular, the bifacial DSSC with PR-190 attained 6.55% for front illumination and 5.01% for back illumination, while that with TD had 6.42% for front illumination and 4.35% for back illumination. Because our process does not require an elevated temperature, we are currently exploring the fabrication of flexible bifacial DSSCs that employ polymeric substrates. In addition, further optimization (e.g., $[Co(bpy)_3]^{3+/2+}$ electrolytes) should result in better bifacial DSSC performance.

Author Contributions: Conceptualization, A.G. and D.X.L.; methodology, A.G., S.K., and D.X.L.; formal analysis, A.G., S.K., B.M., and D.X.L.; investigation, A.G., D.X.L., and J.H., data curation, A.G. and D.X.L.; writing—original draft preparation, A.G., B.M., and J.H.; writing—review and editing, B.M. and J.H.; visualization, A.G. and J.H.; supervision, J.H.; project administration, J.H.; funding acquisition, J.H. All authors have read and agreed to the published version of the manuscript.

Funding: This research was supported by a National Research Foundation of Korea (NRF) grant funded by the Ministry of Science and ICT (MSIT) of Korea for the First-Mover Program for Accelerating Disruptive Technology Development (NF-2018M3C1B99088457), and the Energy Technology Development program through the Korea Institute of Energy Technology Evaluation and Planning (KETEP), funded by the Ministry of Trade, Industry and Energy (MOTIE) of Korea (No. 20193010014740 and No. 20193020010440).

Acknowledgments: We acknowledge the Chung-Ang University Research Scholarship in 2019.

Conflicts of Interest: The authors declare no conflict of interest.

References

1. Fakharuddin, A.; Jose, R.; Brown, T.M.; Fabregat-Santiago, F.; Bisquert, J. A perspective on the production of dye-sensitized solar modules. *Energ. Environ. Sci.* **2014**, *7*, 3952–3981. [CrossRef]
2. Wu, W.; Wang, J.; Zheng, Z.; Hu, Y.; Jin, J.; Zhang, Q.; Hua, J. A strategy to design novel structure photochromic sensitizers for dye-sensitized solar cells. *Sci. Rep.* **2015**, *5*, 8592. [CrossRef] [PubMed]
3. Freitag, M.; Teuscher, J.; Saygili, Y.; Zhang, X.; Giordano, F.; Liska, P.; Hua, J.; Zakeeruddin, S.M.; Moser, J.E.; Grätzel, M.; et al. Dye-sensitized solar cells for efficient power generation under ambient lightning. *Nat. Photonics* **2017**, *11*, 372–378. [CrossRef]
4. Yuan, H.; Wang, W.; Xu, D.; Xu, Q.; Xie, J.; Chen, X.; Zhang, T.; Xiong, C.; He, Y.; Zhang, Y.; et al. Outdoor testing and ageing of dye-sensitized solar cells for building integrated photovoltaics. *Solar Energy* **2018**, *165*, 233–239. [CrossRef]
5. Li, P.; Tang, Q. Highly transparent metal selenide counter electrodes for bifacial dye-sensitized solar cells. *J. Power Sources* **2016**, *317*, 43–48. [CrossRef]
6. Kang, J.S.; Kim, J.; Kim, J.-Y.; Lee, M.J.; Kang, J.; Son, Y.J.; Jeong, J.; Park, S.H.; Ko, M.J.; Sung, Y.-E. Highly efficient bifacial dye-sensitized solar cells employing polymeric counter electrodes. *ACS Appl. Mater. Interfaces* **2018**, *10*, 8611–8620. [CrossRef]

7. Bu, C.; Liu, Y.; Yu, Z.; You, S.; Huang, N.; Liang, L.; Zhao, X.-Z. Highly transparent carbon counter electrode prepared via an in situ carbonization method for bifacial dye-sensitized solar cells. *ACS Appl. Mater. Interfaces* **2013**, *5*, 7432–7438. [CrossRef]
8. Zhang, Y.; Zhao, Y.; Duan, J.; Tang, Q. S-doped CQDs tailored transparent counter electrodes for high-efficiency bifacial dye-sensitized solar cells. *Electrochim. Acta.* **2018**, *216*, 588–595. [CrossRef]
9. Hu, Z.; Xia, K.; Zhang, J.; Hu, Z.; Zhu, Y. Highly transparent ultrathin metal sulfide films as efficient counter electrodes for bifacial dye-sensitized solar cells. *Electrochim. Acta* **2015**, *170*, 39–47. [CrossRef]
10. Bu, C.; Tai, Q.; Liu, Y.; Guo, S.; Zhao, X. A transparent and stable polypyrrole counter electrode for dye-sensitized solar cell. *J. Power Sources* **2013**, *221*, 78–83. [CrossRef]
11. Li, H.; Xiao, X.; Han, G.; Hou, H. Honeycomb-like poly(3,4-ethylenedioxythiophene) as an effective and transparent counter electrode in bifacial dye-sensitized solar cells. *J. Power Sources* **2017**, *342*, 709–716. [CrossRef]
12. He, B.; Zhang, X.; Zhang, H.; Li, J.; Meng, Q.; Tang, Q. Transparent molybdenum sulfide decorated polyaniline complex counter electrodes for efficient bifacial dye-sensitized solar cells. *Solar Energy* **2017**, *147*, 470–478. [CrossRef]
13. Ahmed, A.S.A.; Xiang, W.; Hu, X.; Qi, C.; Amiinu, I.S.; Zhao, X. Si_3N_4/MoS_2-PEDOT:PSS composite counter electrode for bifacial dye-sensitized solar cells. *Solar Energy* **2018**, *173*, 1135–1143. [CrossRef]
14. Moraes, R.S.; Saito, E.; Leite, D.M.G.; Massi, M.; da Silva Sobrinho, A.S. Optical, electrical and electrochemical evaluation of sputtered platinum counter electrodes for dye sensitized solar cells. *Appl. Surf. Sci.* **2016**, *364*, 229–234. [CrossRef]
15. Zhou, R.; Guo, W.; Yu, R.; Pan, C. Highly flexible, conductive and catalytic Pt networks as transparent counter electrodes for wearable dye-sensitized solar cells. *J. Mater. Chem. A* **2015**, *3*, 23028–23034. [CrossRef]
16. Dong, H.; Chen, Y.C.; Feldmann, C. Polyol synthesis of nanoparticles: Status and options regarding metals, oxides, chalcogenides, and non-metal elements. *Green Chem.* **2015**, *17*, 4107–4132. [CrossRef]
17. Fievet, F.; Ammar-Merah, S.; Brayner, R.; Chau, F.; Giraud, M.; Mammeri, F.; Peron, J.; Piquemal, J.Y.; Sicard, L.; Viau, G. The polyol process: A unique method for easy access to metal nanoparticles with tailored sizes, shapes and compositions. *Chem. Soc. Rev.* **2018**, *47*, 5187–5233. [CrossRef]
18. Cho, S.J.; Ouyang, J. Attachment of platinum nanoparticles to substrates by coating and polyol reduction of a platinum precursor. *J. Phys. Chem. C* **2011**, *115*, 8519–8526. [CrossRef]
19. Sun, K.; Fan, B.; Ouyang, J. Nanostructured platinum films deposited by polyol reduction of a platinum precursor and their application as counter electrode dye-sensitized solar cells. *J. Phys. Chem. C* **2010**, *114*, 4237–4244. [CrossRef]
20. Cho, S.J.; Neo, C.Y.; Mei, X.G.; Ouyang, J.Y. Platinum nanoparticles deposited on substrates by solventless chemical reduction of a platinum precursor with ethylene glycol vapor and its application as highly effective electrocatalyst in dye-sensitized solar cells. *Electrochim. Acta* **2012**, *85*, 16–24. [CrossRef]
21. Song, M.Y.; Chaudhari, K.N.; Park, J.; Yang, D.-S.; Kim, J.H.; Kim, M.-S.; Lim, K.; Ko, J.; Yu, J.-S. High efficient Pt counter electrode prepared by homogeneous deposition method for dye-sensitized solar cell. *Appl. Energy* **2012**, *100*, 132–137. [CrossRef]
22. Decoppet, J.-D.; Moehl, T.; Babkair, S.S.; Alzubaydi, R.A.; Ansari, A.A.; Habib, S.S.; Zakeeruddin, S.M.; Schmidt, H.-W.; Grätzel, M. Molecular gelation of ionic liquid–sulfolane mixtures, a solid electrolyte for high performance dye-sensitized solar cells. *J. Mater. Chem. A* **2014**, *2*, 15972–15977. [CrossRef]
23. Leonardi, E.; Penna, S.; Brown, T.M.; Di Carlo, A.; Reale, A. Stability of dye-sensitized solar cells under light soaking test. *J. Non-Cryst. Solids* **2010**, *356*, 2049–2052. [CrossRef]
24. Veerender, P.; Saxena, V.; Jha, P.; Koiry, S.P.; Gusain, A.; Samanta, S.; Chauhan, A.K.; Aswal, D.K.; Gupta, S.K. Free-standing polypyrrole films as substrate-free and Pt-free counter electrodes for quasi-solid dye-sensitized solar cells. *Org. Electron.* **2012**, *13*, 3032–3039.
25. Skrabalak, S.E.; Wiley, B.J.; Kim, M.; Formo, E.V.; Xia, Y. On the polyol synthesis of silver nanostructures: Glycolaldehyde as a reducing agent. *Nano Lett.* **2008**, *8*, 2077–2081. [CrossRef] [PubMed]
26. Dablemont, C.; Lang, P.; Mangeney, C.; Piquemal, J.-Y.; Petkov, V.; Herbst, F.; Viau, G. FTIR and XPS study of Pt nanoparticle functionalization and interaction with alumina. *Langmuir* **2008**, *24*, 5832–5841. [CrossRef]
27. Liu, Z.; Lee, J.Y.; Chen, W.; Han, M.; Gan, L.M. Physical and electrochemical characterizations of microwave-assisted polyol preparation of carbon-supported PtRu nanoparticles. *Langmuir* **2004**, *20*, 181–187. [CrossRef]

28. Wan, J.; Fang, G.; Yin, H.; Liu, X.; Liu, D.; Zhao, M.; Ke, W.; Tao, H.; Tang, Z. Pt-Ni alloy nanoparticles as superior counter electrodes for dye-sensitized solar cells: Experimental and theoretical understanding. *Adv. Mater.* **2014**, *26*, 8101–8106. [CrossRef]
29. Chang, P.-J.; Cheng, K.-Y.; Chou, S.-W.; Shyne, J.-J.; Yang, Y.-Y.; Hung, C.-Y.; Lin, C.-Y.; Chen, H.-L.; Chou, H.-L.; Chou, P.-T. Tri-iodide reduction activity of shape- and composition-controlled PtFe nanostructures as counter electrodes in dye-sensitized solar cells. *Chem. Mater.* **2016**, *28*, 2110–2119. [CrossRef]
30. Popoola, I.K.; Gondal, M.A.; AlGhamdi, J.M.; Qahtan, T.F. Photofabrication of highly transparent platinum counter electrodes at ambient temperature for bifacial dye sensitized solar cells. *Sci. Rep.* **2018**, *8*, 12864.

Article

Arrays of Plasmonic Nanostructures for Absorption Enhancement in Perovskite Thin Films

Tianyi Shen, Qiwen Tan, Zhenghong Dai, Nitin P. Padture and Domenico Pacifici *

School of Engineering, Brown University, 184 Hope Street, Providence, RI 02912, USA; tianyi_shen@brown.edu (T.S.); qiwen_tan@brown.edu (Q.T.); zhenghong_dai@brown.edu (Z.D.); nitin_padture@brown.edu (N.P.P.)

* Correspondence: domenico_pacifici@brown.edu

Received: 15 June 2020; Accepted: 7 July 2020; Published: 9 July 2020

Abstract: We report optical characterization and theoretical simulation of plasmon enhanced methylammonium lead iodide ($MAPbI_3$) thin-film perovskite solar cells. Specifically, various nanohole (NH) and nanodisk (ND) arrays are fabricated on gold/$MAPbI_3$ interfaces. Significant absorption enhancement is observed experimentally in 75 nm and 110 nm-thick perovskite films. As a result of increased light scattering by plasmonic concentrators, the original Fabry–Pérot thin-film cavity effects are suppressed in specific structures. However, thanks to field enhancement caused by plasmonic resonances and in plane interference of propagating surface plasmon polaritons, the calculated overall power conversion efficiency (PCE) of the solar cell is expected to increase by up to 45.5%, compared to its flat counterpart. The role of different geometry parameters of the nanostructure arrays is further investigated using three dimensional (3D) finite-difference time-domain (FDTD) simulations, which makes it possible to identify the physical origin of the absorption enhancement as a function of wavelength and design parameters. These findings demonstrate the potential of plasmonic nanostructures in further enhancing the performance of photovoltaic devices based on thin-film perovskites.

Keywords: perovskite solar cells; surface plasmon polaritons; plasmonic nanostructures; absorption enhancement; FDTD simulations

1. Introduction

Hybrid organic–inorganic perovskites have become one of the most popular photovoltaic materials due to their high absorption coefficient [1], long carrier diffusion length [2] as well as low-cost fabrication process [3–7], which make them excellent candidates for high-efficiency solar cells. As one of the most popular members of perovskite materials, methylammonium lead iodide or $CH_3NH_3PbI_3$ ($MAPbI_3$) possesses such a high absorption coefficient that a 400 nm-thick film is sufficient to absorb most of the incident solar spectrum below its bandgap. Although the record power conversion efficiency (PCE) of a single-junction perovskite solar cell has reached values up to 25.2% [8], there is still much room to boost the PCE. Therefore, several methods have been proposed to improve the performance of perovskite solar cells, one of which involves the use of surface plasmons. Plasmonic effects could be employed to improve the performance of perovskite solar cells [9–15] by embedding nanoparticles [16–30] or plasmonic concentrators [31–35] that can increase absorption, especially near the optical band edge of the material. Previous theoretical research has shown that nanohole (NH) or nanodisk (ND) arrays embedded on gold electrodes with a thin layer of $MAPbI_3$ deposited on them can significantly improve the solar cell PCE by up to ∼10%. The reduced film thickness has the implied benefits of reducing the amount of toxic materials, improving electronic performance and enabling fabrication on flexible substrates [36].

In this work, we report the experimental fabrication, structural and optical characterization, as well as theoretical simulations of plasmon enhanced $MAPbI_3$ perovskite solar cell with NH and ND arrays on gold substrates, as a function of geometry parameters and varying thickness of $MAPbI_3$ deposited on top. Four geometry features, height h, diameter D, pitch P and $MAPbI_3$ thickness t, are systematically studied by three-dimensional finite-difference time-domain (3D FDTD) simulations to identify the physical effects responsible for the absorption and PCE enhancement.

2. Materials and Methods

An adhesion layer of 4 nm-thick titanium and a layer of 200 nm-thick gold was sequentially deposited on a 1 mm-thick quartz substrate using electron beam evaporation. Subsequently, NH or ND arrays of varying geometry parameters were fabricated on the thick gold film using focused ion beam (FIB) milling.

Following the FIB milling, the substrates were treated with UV-ozone for 45 min to enhance the wettability. The $MAPbI_3$ precursor solution was prepared by dissolving 159 mg of methylammonium iodide (Greatcell, Queanbeyan, Australia) and 461 mg of PbI_2 (Sigma-Aldrich, St. Louis, MO, USA) in 78 mg dimethyl sulfoxide (Sigma-Aldrich, St. Louis, MO, USA) and 1368 mg of N,N-dimethylformamide (Acros organics, NJ, USA) to obtain a 30 wt% solution. To deposit the $MAPbI_3$ layer, the solution was spin-coated at 4000 rpm for 30 s with an acceleration of 1300 rpm/s in a nitrogen-filled glove box. At the 10th second of spinning, 250 μL diethyl ether (Sigma-Aldrich, St. Louis, MO, USA) was dripped onto the substrate. The as-coated film was then annealed at 100 °C for 20 min to obtain the $MAPbI_3$ thin film. The film thickness can be controlled by adjusting the amount of N,N-dimethylformamide. Schematic illustrations of the fabricated NH and ND array coated with $MAPbI_3$ thin film are displayed in Figure 1a,b, respectively.

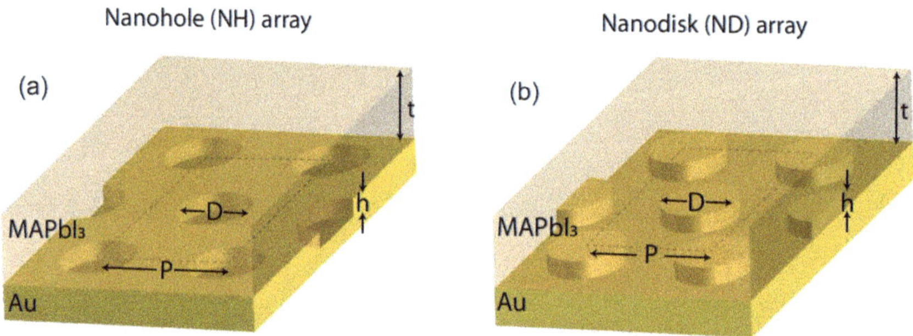

Figure 1. Schematic illustrations of methylammonium lead iodide ($MAPbI_3$) thin-film over nanostructured gold (Au) surface with (**a**) NH or (**b**) ND array, respectively. t represents the perovskite film thickness on the top of gold surface. P represents the triangular array pitch, while h and D respectively represent the height and diameter of a plasmonic concentrator. A negative h indicated NH, while a positive h corresponded to ND.

The thickness of the synthesized $MAPbI_3$ films were characterized using a variable angle spectroscopic ellipsometer (J.M. Woollam, Lincoln, NE, USA, M-2000). The reflectance spectra were characterized using an inverted microscope (Nikon Instruments, Melville, NY, USA, Eclipse Ti) with a small numerical aperture objective (10×, NA = 0.3) to mimic normal incidence. A randomly polarized broadband light source was used to illuminate the sample. The reflected light from the samples was coupled into a spectrograph (Princeton Instruments, Acton, MA, USA, Acton SpectraPro SP-2300) and was detected using a charge-coupled device camera (Princeton Instruments, Acton, MA, USA, Pixis 100). A piece of flat silicon wafer was used as a calibration reference during the characterization.

Apart from experimentally characterizing the absorptance spectra, simulations are also performed using FDTD method. A commerical-grade 3D electromagnetic simulator from Lumerical Inc.

(Vancouver, BC, Canada) was used [37]. Anti-symmetric and symmetric boundary conditions were used in the lateral directions while perfectly matched layer boundary conditions were used in the vertical directions. The real and imaginary parts of refractive indices for gold and $MAPbI_3$ were obtained from literature [38,39]. Considering the balance between computational cost and accuracy for different structures, the mesh size varied among 4 nm × 4 nm × 2 nm, 4 nm × 4 nm × 1 nm and 3 nm × 3 nm × 2 nm in the Cartesian coordinates.

3. Results and Discussion

Figure 2a,b,e,f show the tilted and top-view scanning electron microscope (SEM) micrographs of example NH and ND arrays prior to the $MAPbI_3$ synthesis. To investigate the absorption enhancement brought by the plasmonic nanostructure, the corresponding reflectance spectra $R(\lambda)$ were measured, and the measured absorptance spectra $A(\lambda) = 1 - R(\lambda)$ are reported in Figure 2c,g. Absorptance of nanostructured gold surface (black solid lines) shows a similar spectral shape as that of flat gold surface (black dashed lines) for both NH and ND arrays, while the NH array exhibits relatively higher absorption. This is because a portion of the scattered light by plasmonic concentrators is not collected by the finite NA of the objective lens, therefore a lower measured reflectance (i.e., higher absorption). By contrast, ND array in Figure 2e,f has smaller feature size and larger array pitch. Therefore, a smaller portion of light is scattered and a higher portion of reflected light is collected by the objective (i.e., lower absorption). The scattered light can be efficiently absorbed by adding a layer of $MAPbI_3$ on the top. After being coated with a 110 nm-thick $MAPbI_3$ film, the absorptance spectrum of the flat-interface structure (red dashed lines) shows a peak at wavelength λ = 526 nm. This results from the Fabry–Pérot cavity effect supported by the perovskite thin-film. The absorption decreases with increasing wavelength because the perovskite layer no longer supports any Fabry–Pérot cavity mode and the material has weaker absorbing property at longer wavelength. With NH or ND nanostructures (red solid lines), the absorption gets significantly enhanced at longer wavelengths, as is shown as red fillings in Figure 2d,h. For NH array, the spectral peak at λ = 526 nm gets weaker. This is because the structure disrupts the interference in the film [36]. In contrast, ND array still preserves that spectral peak due to its smaller feature size and lower nanostructure concentration. The absorptance spectra of perovskite/nanostructured gold are normalized to that of perovskite/flat gold and are displayed in Figure 2d,h. The blue filling at lower wavelengths represents the suppression of Fabry–Pérot cavity mode while the red filling represents the absorption enhancement brought by the plasmonic nanostructure. It is obvious that the example NH could not preserve the cavity effect as well as ND array (larger blue filling area), but the greater absorption enhancement by the plasmonic effects at longer wavelengths compensates for the loss at lower wavelengths (larger red filling area). As a comparison, for the example ND array, both blue and red filling areas are much smaller in Figure 2h, indicating negligible suppression at short wavelengths but also less significant enhancement at longer wavelengths.

A variety of nanostructured gold surfaces (NH or ND arrays) were fabricated based on optimal design considerations [36]. The top row of Figure 3 shows the SEM micrographs of the NH (left) and ND arrays (right). The measured and normalized absorptance spectra of these NH and ND arrays are shown in Figure 3a–d. For NH arrays (left column), as pitch P increases from 200 nm (circle symbols) to 250 nm (triangle symbols), the spectral peaks redshift from λ = 632 nm to λ = 666 nm. This results from the pitch-dependent collective surface plasmon polariton (SPP) resonances in the hexagonal lattice [36,40]. In addition, with hole depth increasing from 40 nm (solid symbols) to 70 nm (open symbols), the Fabry–Pérot cavity mode gets disrupted, resulting in a relatively lower absorption at λ = 526 nm, but there is more significant absorption enhancement at longer wavelengths. A similar behavior is also observed in ND arrays: structures suppressing the Fabry–Pérot thin-film interference effects show stronger absorption enhancement near the $MAPbI_3$ band edge.

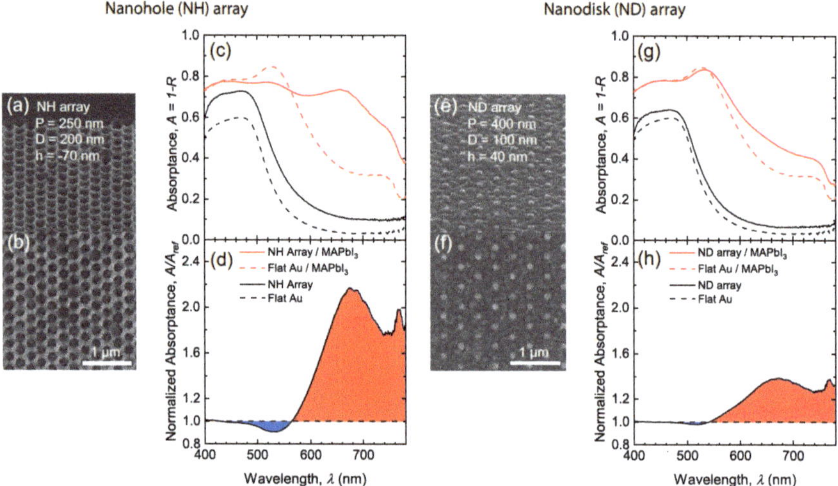

Figure 2. (**a**,**b**,**e**,**f**) display the tilted-view (**a**,**e**) and top-view (**b**,**f**) SEM micrographs of the fabricated nanohole (NH) array with P = 250 nm, D = 200 nm, h = −70 nm (**a**,**b**) and nanodisk (ND) array with P = 400 nm, D = 100 nm, h = 40 nm (**e**,**f**). (**c**,**g**) The measured absorptance spectra $A = 1 - R$ for nanostructured (solid lines) or flat (dashed lines) gold with (red lines) or without (black lines) 110nm-thick $MAPbI_3$ thin-film. (**d**,**h**) show the absorptance spectra for 110nm-thick $MAPbI_3$ thin-films on nanostructured gold normalized to that of equal-thickness $MAPbI_3$ on flat gold. The red filling ($A/A_{ref} > 1.0$) represents absorption enhancement while the blue filling ($A/A_{ref} < 1.0$) represents absorption suppression.

The absorption enhancement with different perovskite layer thicknesses is also investigated experimentally. $MAPbI_3$ layers with different thicknesses (t = 75, 110, 300 nm) were synthesized on the same nanostructure and the corresponding absorptance spectra were characterized and displayed in Figure 4. The spectral peaks and dips in the flat-interface cases (dashed lines) again result from the Fabry–Pérot cavity modes supported by different perovskite thicknesses at different wavelengths. The SEM micrographs of nanostructured surfaces are displayed in the top panels. Low-density and small-feature-size nanostructures and the corresponding absorptance spectra are shown in the left column. For different thicknesses t, these two structures can preserve the Fabry–Pérot cavity modes, as well as bring additional absorption enhancement. However, the absorption enhancement becomes much less significant as t increases to 300 nm, because most of the incident light is absorbed by the perovskite material before interacting with the plasmonic concentrators. For NH or ND arrays with larger feature sizes and higher concentration (right column), Fabry–Pérot cavity effects are significantly suppressed at the resonant wavelengths. In addition, both high-density and large-feature-size arrays exhibit stronger absorption enhancement at other wavelength ranges comparing to the low-density and small-feature-size arrays. Moreover, obvious absorption enhancement is observed even when t increases to 300 nm.

To further evaluate how the plasmonic effects and absorption enhancement will affect the overall performance of real photovoltaic devices, the PCE can be obtained using the measured absorptance spectra for different active layer thicknesses. With the detailed balance assumption, PCE can be expressed as

$$\mathrm{PCE} = \frac{\int_{\lambda<\lambda_g} \mathrm{AM1.5}(\lambda)\frac{2\pi\lambda}{\hbar c}A(\lambda)E_g d\lambda}{1\,sun} \tag{1}$$

where λ_g is the bandgap wavelength of $MAPbI_3$, AM1.5(λ) is the wavelength-dependent air mass 1.5 solar radiation spectrum, $\hbar$ is the reduced Planck constant and 1 sun is the 1000 W/m^2 incident solar power density [41]. The experimentally measured absorptance spectra includes the absorption in both gold and $MAPbI_3$, but the portion absorbed by gold is relatively small according to previously reported study [36]. For NH and ND arrays in Figure 4 coated with multiple active layer thicknesses, the calculated PCE and PCE enhancement is shown in Figure 5.

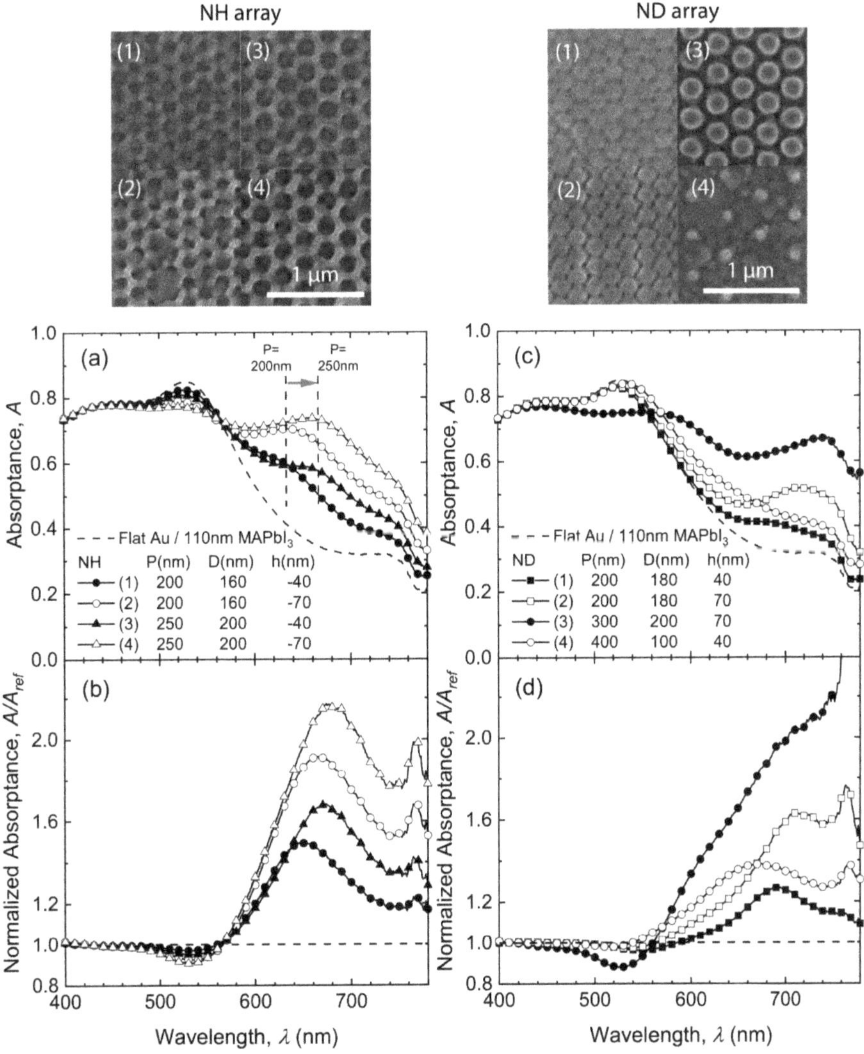

Figure 3. Top-view SEM micrographs (top), experimental absorptance spectra (**a**,**c**) and normalized absorptance spectra (**b**,**d**) for 110nm-thick $MAPbI_3$ on flat (dashed lines) and nanostructured (solid lines with symbols) gold NH (left column) and ND (right column) arrays. The vertical dashed lines in (**a**) show the absorption spectral redshift as a function of pitch *P* (increasing from 200 nm to 250 nm).

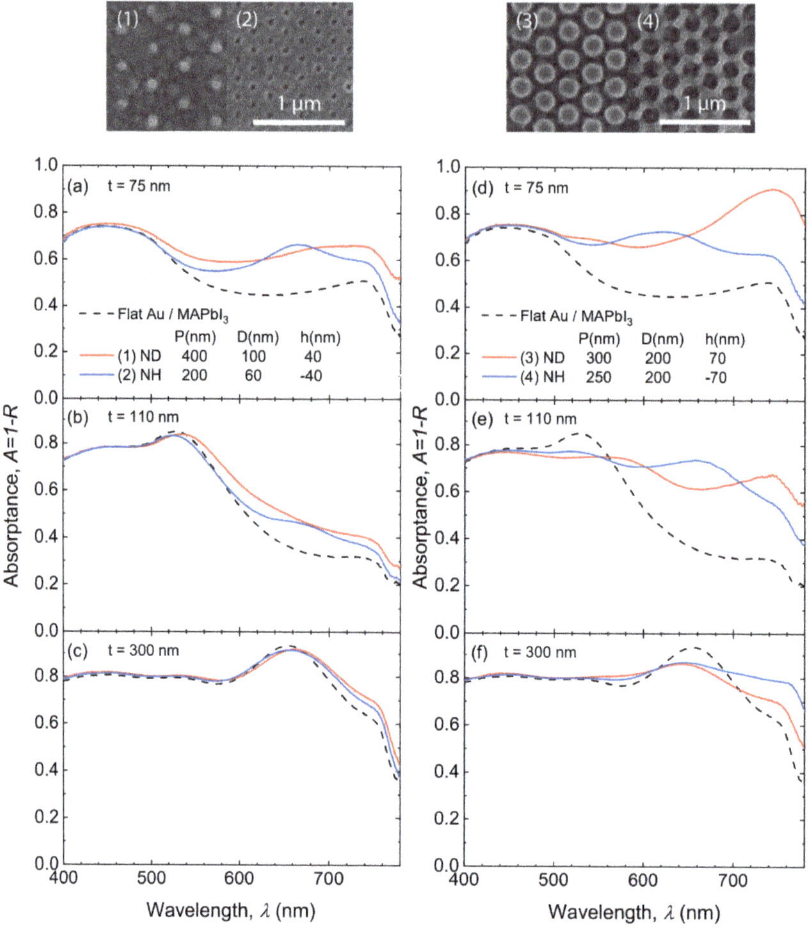

Figure 4. Measured absorptance spectra of flat (dashed lines) and nanostructured (solid lines) gold surfaces coated with varying-thickness $MAPbI_3$ with t = 75 nm (**a**,**d**); 110 nm (**b**,**e**); 300 nm (**c**,**f**). The left (right) column corresponds to nanostructured gold surfaces with low (high) surface density of optical scatterers. The top panels show top-view SEM micrographs of gold surfaces with low (1–2) and high-concentration (3–4) ND or NH arrays.

The PCE of flat structure $PCE_{flat}(t)$ (dashed line) is obtained using the multilayer interference model. For all three thicknesses (t = 75, 110 and 300 nm), all four nanostructure arrays exhibit PCE enhancement. Specifically, for active layer thickness t = 75 nm, high-density NH and ND arrays (red symbols) increase the PCE by 26.6% and 45.5%, respectively.

As a comparison, low-density nanostructure arrays (blue symbols) do not enhance the PCE as much. As the thickness increases to 300 nm, the PCE enhancement becomes much less significant due to the weaker plasmonic effect. The performance difference narrows between arrays of different concentration and feature sizes, enhancing the PCE by 3.7–7.0%.

As demonstrated in the experiment, plasmonic nanostructures with various geometry parameters exhibit different absorption enhancement behavior. Therefore, FDTD simulations were performed to further investigate the influence of different geometry parameters on the absorption enhancement.

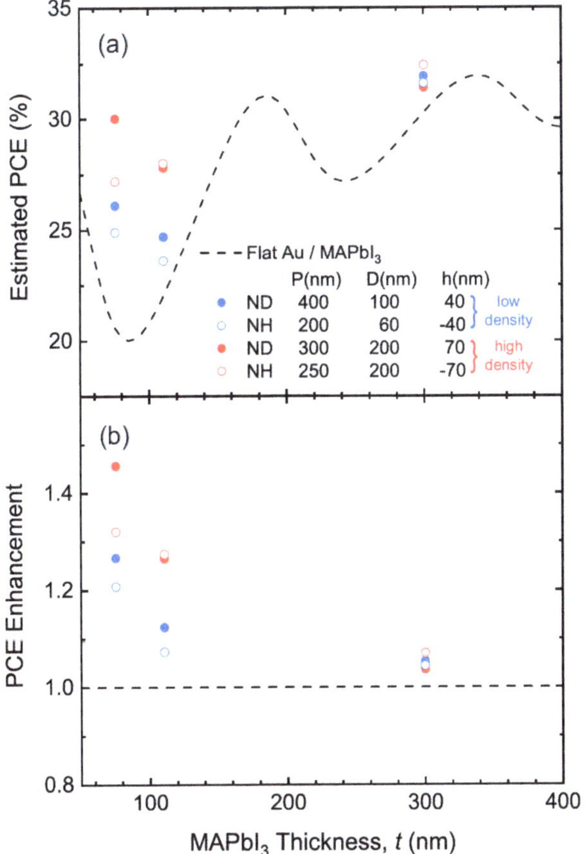

Figure 5. (**a**) Power conversion efficiency (PCE) and (**b**) PCE enhancement estimated from experimental absorptance spectra for varying $MAPbI_3$ thickness values, t = 75, 110 and 300 nm with low (blue symbols) and high (red symbols) surface density of optical scatterers in NH (open symbols) and ND arrays (solid symbols). The dashed lines represent the PCE of flat-interface structures as a function of $MAPbI_3$ thickness. A strong PCE enhancement is clearly demonstrated for high-density nanostructures at t = 75 and 110 nm (red symbols).

Figure 6a–d display the simulated absorptance spectra of an example NH array (P = 250 nm, D = 200 nm, h = −70 nm) coated with 110 nm-thick $MAPbI_3$ film. In the left column, each of the four parameters are individually varied, meanwhile the other three parameters are kept fixed. When each parameter is varied, specific spectra with spectral peaks of interest are selected and respectively exhibited in Figure 6e–h and the insets show the electric field intensity distribution at the peak wavelengths.

First, the role of NH depth $|h|$ is investigated. It is noticeable that the spectral peak at λ = 545 nm redshifts when $|h|$ increases from 10 nm (black solid line) to 70 nm (yellow line). This peak results from the shallow hole, not disrupting the Fabry–Pérot effect (the bright horizontal band in inset (1)). When $|h|$ increases to 50 nm, the peak redshifts due to a localized surface plasmon resonance (LSPR) arising inside the hole (inset (2) in Figure 6e). When $|h|$ further increases to 150 nm, the constructive interference near the $MAPbI_3$/gold interface is restored, but is not as strong (faint horizontal band in inset (3) in Figure 6e).

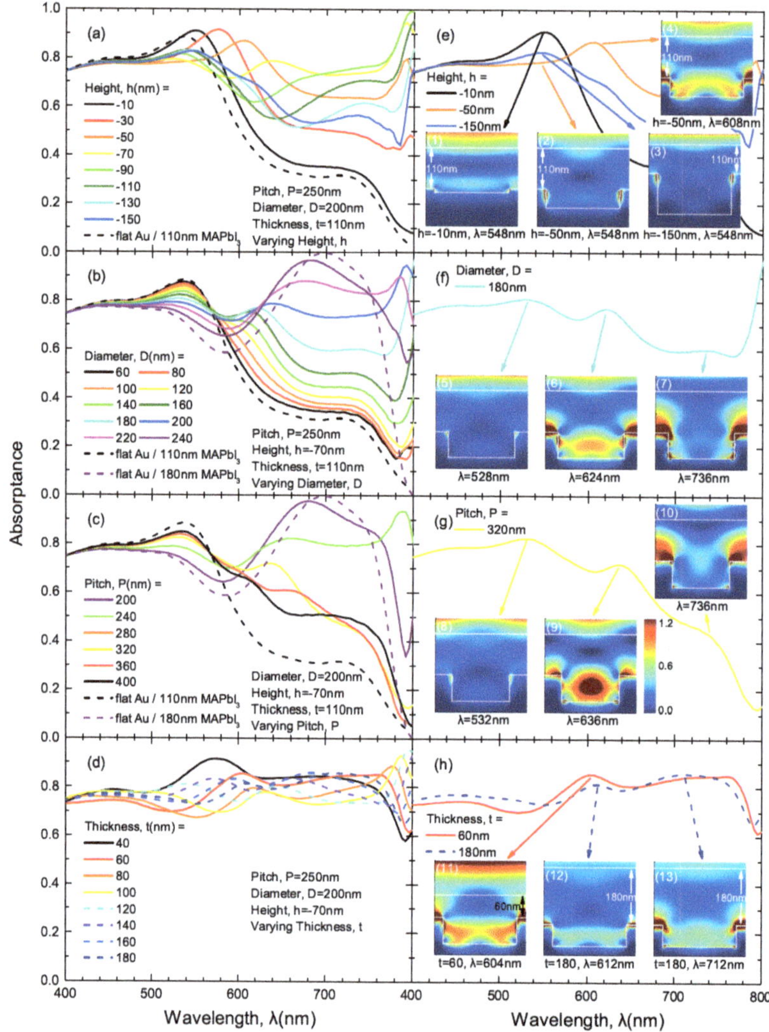

Figure 6. (**a**–**d**) Simulated absorptance spectra of representative NH arrays coated with a 110 nm-thick $MAPbI_3$ film. Each of the geometry parameters (i.e., height, *h*, diameter, *D*, pitch, *P* and $MAPbI_3$ thickness, *t*) is individually varied while keeping the others constant. (**e**–**h**) Representative absorptance spectra from (**a**–**d**) showing spectral peaks of interest. The color maps reported as insets show cross sections of electric field intensity simulated at each corresponding spectral peak wavelength by using 3D finite-difference time-domain (FDTD).

Figure 6b,c respectively demonstrate the absorptance spectra change with varying pitch *P* and diameter *D*. The black and purple dashed lines correspond to absorptance spectra of 110 nm and (110+$|h|$) nm-thick $MAPbI_3$ film on flat gold, respectively. With small feature sizes (small diameter *D*) or low NH concentration (large pitch *P*), the absorptance spectra are similar to that of 110 nm-thick $MAPbI_3$ on flat gold (black dashed lines). As the NH size or concentration increases, the absorptance spectral shape starts to shift towards that of 180 nm-thick $MAPbI_3$ on flat gold (purple dashed lines), which results from the Fabry–Pérot mode between the bottom of the holes and top of $MAPbI_3$ film. In addition, another peak near 630 nm shows up in some structures. Insets (6, 9) in Figure 6f,g show that these spectral peaks attribute to the LSPR inside the NH structure. Moreover, an additional spectral peak is observed at λ = 736 nm. These peaks results from the LSPR between neighboring NH structures, as is shown in insets (7, 10) in Figure 6f,g. Figure 6d demonstrates the thickness dependence of the absorptance spectra. With increasing thickness, the multiple peaks corresponds to the LSPR inside the holes or between neighboring holes. In addition, the resonances gets weaker as t increases due to less light–matter interaction.

Likewise, the role of the ND geometry parameters in absorption enhancement is also investigated. Figure 7 shows the simulated absorptance spectra of an example ND array (P = 400 nm, D = 100 nm, h = 40 nm) coated with $MAPbI_3$ thin film of t = 110 nm.

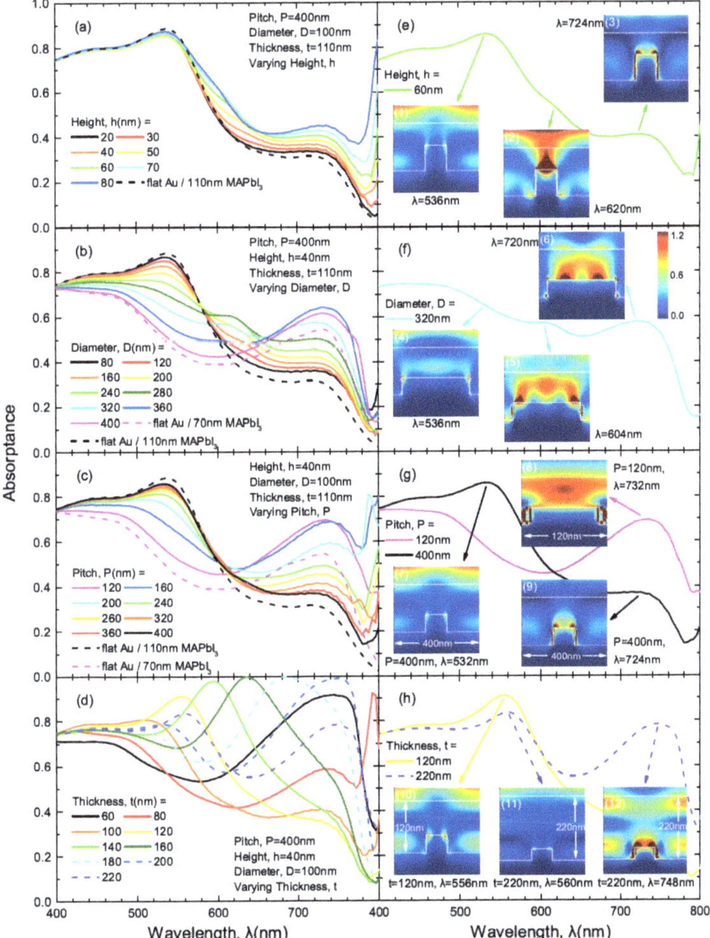

Figure 7. (**a**–**d**) Simulated absorptance spectra of representative ND arrays coated with a 110 nm-thick $MAPbI_3$ film. The geometry parameters (i.e., height, h, diameter, D, pitch, P and $MAPbI_3$ thickness, t) are individually varied while the other three parameters are kept constant; (**e**–**h**) Representative absorptance spectra from (**a**–**d**) showing spectral peaks of interest. The color maps reported as insets show cross sections of electric field intensity simulated at each corresponding spectral peak wavelength by using 3D FDTD.

While varying h (Figure 7a), the absorptance spectra shape mainly follows that of flat-interface structure (dashed black line), showing peaks at λ = 530 nm, which corresponds to the Fabry–Pérot effect of 110 nm-thick $MAPbI_3$ on gold. Insets (1, 3) in Figure 7e show a discontinuous horizontal band, indicating that the existence of ND disrupts the cavity effect, but the spectral peaks still persists. Additionally, the ND also brings additional plasmonic resonances, as is shown in inset (3), enhancing the absorption with increasing height. Moreover, for a specific structure (h = 60 nm), another resonance is observed on the top of the disk structure, as is shown in inset (2) in Figure 7e.

The diameter D and array pitch P play a similar role in ND arrays: smaller feature sizes and low concentration preserve cavity effects, but also contribute to less plasmonic effect, as is shown in Figure 7b,c. For D = 320 nm (Figure 7f), specifically, the suppression of Fabry–Pérot effect is observed (inset (4) in Figure 7f). Similar to NH arrays, high-density (small P) ND arrays also suppress the cavity effects, while contributing to significant plasmonic effects (insets (7–9) in Figure 7g). In addition, multiple orders of localized standing-wave surface plasmonic resonances are observed on the top surface of the ND structure, as are reported in insets (5, 6) in Figure 7f.

Figure 7d shows the absorptance spectra of structures with varying $MAPbI_3$ thickness t coated on the ND arrays. Due to the small ND size, as well as the relatively large array pitch, the spectra peaks generally result from the Fabry–Pérot cavity effects (horizontal bands in insets (10–12) in Figure 7h). In spite of the existence of the ND structure, the modes still persist.

The PCE is calculated based on the previous simulation results, and is reported as black circle lines in Figure 8. The PCE of NH and ND structure are respectively displayed in the left and right column. As is expected, when coated with 110nm-thick $MAPbI_3$, most of the structures with nanostructured surfaces exhibit significantly enhanced PCE (black circles) compared to their flat-interface counterparts (red dashed lines). It is worth noting that NH structures are adding extra perovskite material volume to the 110 nm-thick film, while ND structures have less perovskite material than the flat structure. To take this into account, two additional references are exhibited in Figure 8: (1) the blue dashed lines in Figure 8a–c and gray dashed lines in Figure 8e–g respectively show $PCE_{flat}(t = 110 \pm |h| \text{ nm})$ for NH and ND arrays; (2) the red triangle lines represent the PCE of flat structures with a combined thickness that accounts for the presence of two effective Fabry–Pérot cavities, which can be respectively expressed as

$$PCE_{combined} = \frac{\pi D^2}{2\sqrt{3}P^2} PCE_{flat}(\text{t=110} \pm |h|\text{nm}) + (1 - \frac{\pi D^2}{2\sqrt{3}P^2}) PCE_{flat}(\text{t=110nm}) \quad (2)$$

where the plus (minus) sign refers to NH (ND) arrays. For most of the structures, the nanostructured-surface PCE is greater than $PCE_{combined}$, which further proves that the plasmonic effects improve the solar cell performance. In addition, the difference between these two PCEs generally increases with increasing D and decreasing P, i.e., the cell performance benefits more from high-density and large-size nanostructures. In addition, with small thickness t, a larger portion of light penetrates the perovskite active layer, interacting strongly with the nanostructure array, and therefore also results in a larger difference.

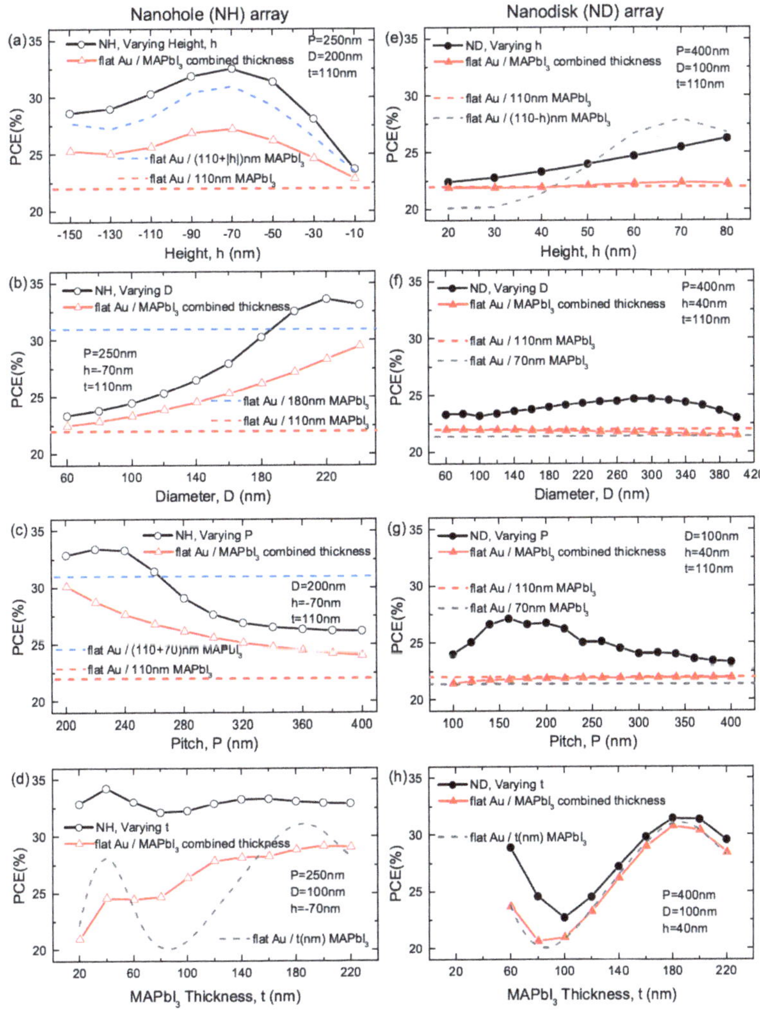

Figure 8. Calculated PCE for NH (**a**–**d**) and ND (**e**–**h**) arrays as a function of varying geometry parameters (i.e., height, *h*, diameter, *D*, array pitch, *P* and $MAPbI_3$ thickness, *t*). Dashed lines show the PCE of flat structures with $t = 110$ nm or $110 \pm |h|$ nm, while red triangle lines show the PCE of flat structures with a combined thickness that accounts for the presence of two effective Fabry–Pérot cavities in the nanostructured structure.

4. Conclusions

In conclusion, various plasmonic nanostructure arrays are experimentally fabricated on gold surfaces coated with perovskite films with varying thickness and their optical absorption properties are theoretically calculated by using 3D FDTD simulations. It is shown that, although thin-film interference effects can be negatively affected by the presence of nanoscatterers, overall, the plasmonic concentrators allow for significant absorption enhancement as the result of a combination of physical effects, i.e., (1) localized optical resonances (including surface plasmon resonances); (2) plasmonic modes in vertical cavities (such as the space within nanoholes); (3) in-plane constructive interference of surface plasmon polaritons that propagate along the metal/dielectric interface. Structures with higher surface density of nanoscattereres and with reduced film thickness generally show larger absorption and calculated PCE enhancements. The experimental results are supported by 3D FDTD simulations that help identify the individual physical effects and elucidate their interplay in redistributing the incident field intensity, which in turns determine the observed absorption and the calculated PCE enhancements.

Author Contributions: Data curation, T.S. and Q.T.; formal analysis, T.S. and Q.T.; resources, Z.D.; supervision, N.P.P. and D.P.; writing—original draft, T.S. and Q.T.; writing—review and editing, Z.D., N.P.P. and D.P. All authors have read and agreed to the published version of the manuscript.

Funding: This material is based upon work supported by the National Science Foundation under Grant No. OIA-1538893.

Acknowledgments: The authors thank Min Chen for fruitful discussions.

Conflicts of Interest: The authors declare no conflict of interest.

References

1. Huang, J.; Yuan, Y.; Shao, Y.; Yan, Y. Understanding the physical properties of hybrid perovskites for photovoltaic applications. *Nat. Rev. Mater.* **2017**, *2*, 1–19. [CrossRef]
2. Dong, Q.; Fang, Y.; Shao, Y.; Mulligan, P.; Qiu, J.; Cao, L.; Huang, J. Electron-hole diffusion lengths > 175 μm in solution-grown $CH_3NH_3PbI_3$ single crystals. *Science* **2015**, *347*, 967–970. [CrossRef] [PubMed]
3. Yang, S.; Fu, W.; Zhang, Z.; Chen, H.; Li, C.Z. Recent advances in perovskite solar cells: efficiency, stability and lead-free perovskite. *J. Mater. Chem. A* **2017**, *5*, 11462–11482. [CrossRef]
4. Elumalai, N.K.; Mahmud, M.A.; Wang, D.; Uddin, A. Perovskite solar cells: Progress and advancements. *Energies* **2016**, *9*, 861. [CrossRef]
5. Chen, M.; Ju, M.G.; Carl, A.D.; Zong, Y.; Grimm, R.L.; Gu, J.; Zeng, X.C.; Zhou, Y.; Padture, N.P. Cesium titanium (IV) bromide thin films based stable lead-free perovskite solar cells. *Joule* **2018**, *2*, 558–570. [CrossRef]
6. Song, Z.; Watthage, S.C.; Phillips, A.B.; Heben, M.J. Pathways toward high-performance perovskite solar cells: review of recent advances in organo-metal halide perovskites for photovoltaic applications. *J. Photon. Energy* **2016**, *6*, 022001. [CrossRef]
7. Zhou, Y.; Zhou, Z.; Chen, M.; Zong, Y.; Huang, J.; Pang, S.; Padture, N.P. Doping and alloying for improved perovskite solar cells. *J. Mater. Chem. A* **2016**, *4*, 17623–17635. [CrossRef]
8. Best Research–Cell Efficiencies. Available online: https://www.nrel.gov/pv/assets/pdfs/best-research-cell-efficiencies.20200406.pdf (accessed on 8 July 2020).
9. Ferry, V.E.; Sweatlock, L.A.; Pacifici, D.; Atwater, H.A. Plasmonic nanostructure design for efficient light coupling into solar cells. *Nano Lett.* **2008**, *8*, 4391–4397. [CrossRef] [PubMed]
10. Atwater, H.A.; Polman, A. Plasmonics for improved photovoltaic devices. In *Materials For Sustainable Energy: A Collection of Peer-Reviewed Research and Review Articles from Nature Publishing Group*; World Scientific: Singapore, 2011; pp. 1–11.
11. Polman, A.; Atwater, H.A. Photonic design principles for ultrahigh-efficiency photovoltaics. *Nat. Mater.* **2012**, *11*, 174–177. [CrossRef] [PubMed]
12. Gan, Q.; Bartoli, F.J.; Kafafi, Z.H. Plasmonic-enhanced organic photovoltaics: Breaking the 10% efficiency barrier. *Adv. Mater.* **2013**, *25*, 2385–2396. [CrossRef]
13. Yang, J.; You, J.; Chen, C.C.; Hsu, W.C.; Tan, H.r.; Zhang, X.W.; Hong, Z.; Yang, Y. Plasmonic polymer tandem solar cell. *ACS Nano* **2011**, *5*, 6210–6217. [CrossRef] [PubMed]
14. Ding, I.K.; Zhu, J.; Cai, W.; Moon, S.J.; Cai, N.; Wang, P.; Zakeeruddin, S.M.; Grätzel, M.; Brongersma, M.L.; Cui, Y.; et al. Plasmonic dye-sensitized solar cells. *Adv. Energy Mater.* **2011**, *1*, 52–57. [CrossRef]
15. Spinelli, P.; Ferry, V.; Van de Groep, J.; Van Lare, M.; Verschuuren, M.; Schropp, R.; Atwater, H.; Polman, A. Plasmonic light trapping in thin-film Si solar cells. *J. Opt.* **2012**, *14*, 024002. [CrossRef]
16. Jiménez-Solano, A.; Carretero-Palacios, S.; Míguez, H. Absorption enhancement in methylammonium lead iodide perovskite solar cells with embedded arrays of dielectric particles. *Opt. Express* **2018**, *26*, A865–A878. [CrossRef] [PubMed]
17. Miranda-Muñoz, J.M.; Carretero-Palacios, S.; Jiménez-Solano, A.; Li, Y.; Lozano, G.; Míguez, H. Efficient bifacial dye-sensitized solar cells through disorder by design. *J. Mater. Chem. A* **2016**, *4*, 1953–1961. [CrossRef]

18. Carretero-Palacios, S.; Jiménez-Solano, A.; Míguez, H. Plasmonic nanoparticles as light-harvesting enhancers in perovskite solar cells: a user's guide. *ACS Energy Lett.* **2016**, *1*, 323–331. [CrossRef]
19. Carretero-Palacios, S.; Calvo, M.E.; Míguez, H. Absorption enhancement in organic–inorganic halide perovskite films with embedded plasmonic gold nanoparticles. *J. Phys. Chem. C* **2015**, *119*, 18635–18640. [CrossRef]
20. Kang, S.M.; Jang, S.; Lee, J.K.; Yoon, J.; Yoo, D.E.; Lee, J.W.; Choi, M.; Park, N.G. Moth-Eye TiO_2 Layer for Improving Light Harvesting Efficiency in Perovskite Solar Cells. *Small* **2016**, *12*, 2443–2449. [CrossRef]
21. Mali, S.S.; Shim, C.S.; Kim, H.; Patil, P.S.; Hong, C.K. In situ processed gold nanoparticle-embedded TiO_2 nanofibers enabling plasmonic perovskite solar cells to exceed 14% conversion efficiency. *Nanoscale* **2016**, *8*, 2664–2677. [CrossRef]
22. Ghahremanirad, E.; Olyaee, S.; Hedayati, M. The Influence of Embedded Plasmonic Nanostructures on the Optical Absorption of Perovskite Solar Cells. *Photonics* **2019**, *6*, 37. [CrossRef]
23. Mohsen, A.A.; Zahran, M.; Habib, S.; Allam, N.K. Refractory plasmonics enabling 20% efficient lead-free perovskite solar cells. *Sci. Rep.* **2020**, *10*, 1–12. [CrossRef] [PubMed]
24. Deng, W.; Yuan, Z.; Liu, S.; Yang, Z.; Li, J.; Wang, E.; Wang, X.; Li, J. Plasmonic enhancement for high-efficiency planar heterojunction perovskite solar cells. *J. Power Sour.* **2019**, *432*, 112–118. [CrossRef]
25. Wang, B.; Zhu, X.; Li, S.; Chen, M.; Liu, N.; Yang, H.; Ran, M.; Lu, H.; Yang, Y. Enhancing the Photovoltaic Performance of Perovskite Solar Cells Using Plasmonic Au@ Pt@ Au Core-Shell Nanoparticles. *Nanomaterials* **2019**, *9*, 1263. [CrossRef]
26. Yao, K.; Zhong, H.; Liu, Z.; Xiong, M.; Leng, S.; Zhang, J.; Xu, Y.X.; Wang, W.; Zhou, L.; Huang, H.; et al. Plasmonic Metal Nanoparticles with Core–Bishell Structure for High-Performance Organic and Perovskite Solar Cells. *ACS Nano* **2019**, *13*, 5397–5409. [CrossRef] [PubMed]
27. Chen, L.C.; Tien, C.H.; Lee, K.L.; Kao, Y.T. Efficiency Improvement of $MAPbI_3$ Perovskite Solar Cells Based on a $CsPbBr_3$ Quantum Dot/Au Nanoparticle Composite Plasmonic Light-Harvesting Layer. *Energies* **2020**, *13*, 1471. [CrossRef]
28. Lu, Z.; Pan, X.; Ma, Y.; Li, Y.; Zheng, L.; Zhang, D.; Xu, Q.; Chen, Z.; Wang, S.; Qu, B.; et al. Plasmonic-enhanced perovskite solar cells using alloy popcorn nanoparticles. *RSC Adv.* **2015**, *5*, 11175–11179. [CrossRef]
29. Zhang, W.; Saliba, M.; Stranks, S.D.; Sun, Y.; Shi, X.; Wiesner, U.; Snaith, H.J. Enhancement of perovskite-based solar cells employing core–shell metal nanoparticles. *Nano Lett.* **2013**, *13*, 4505–4510. [CrossRef]
30. Bayles, A.; Carretero-Palacios, S.; Calió, L.; Lozano, G.; Calvo, M.E.; Míguez, H. Localized surface plasmon effects on the photophysics of perovskite thin films embedding metal nanoparticles. *J. Mater. Chem. C* **2020**, *8*, 916–921. [CrossRef]
31. Ostfeld, A.; Pacifici, D. Plasmonic concentrators for enhanced light absorption in ultrathin film organic photovoltaics. *Appl. Phys. Lett.* **2011**, *98*, 113112. [CrossRef]
32. Mao, J.; Sha, W.E.; Zhang, H.; Ren, X.; Zhuang, J.; Roy, V.A.; Wong, K.S.; Choy, W.C. Novel direct nanopatterning approach to fabricate periodically nanostructured perovskite for optoelectronic applications. *Adv. Funct. Mater.* **2017**, *27*, 1606525. [CrossRef]
33. Wang, H.; Haroldson, R.; Balachandran, B.; Zakhidov, A.; Sohal, S.; Chan, J.Y.; Zakhidov, A.; Hu, W. Nanoimprinted perovskite nanograting photodetector with improved efficiency. *ACS Nano* **2016**, *10*, 10921–10928. [CrossRef] [PubMed]
34. Long, M.; Chen, Z.; Zhang, T.; Xiao, Y.; Zeng, X.; Chen, J.; Yan, K.; Xu, J. Ultrathin efficient perovskite solar cells employing a periodic structure of a composite hole conductor for elevated plasmonic light harvesting and hole collection. *Nanoscale* **2016**, *8*, 6290–6299. [CrossRef] [PubMed]
35. Paetzold, U.W.; Qiu, W.; Finger, F.; Poortmans, J.; Cheyns, D. Nanophotonic front electrodes for perovskite solar cells. *Appl. Phys. Lett.* **2015**, *106*, 173101. [CrossRef]
36. Shen, T.; Siontas, S.; Pacifici, D. Plasmon-enhanced thin-film perovskite solar cells. *J. Phys. Chem. C* **2018**, *122*, 23691–23697. [CrossRef]
37. Lumerical Inc. Available online: https://www.lumerical.com/products/ (accessed on 8 July 2020).
38. Löper, P.; Stuckelberger, M.; Niesen, B.; Werner, J.; Filipic, M.; Moon, S.J.; Yum, J.H.; Topič, M.; De Wolf, S.; Ballif, C. Complex refractive index spectra of $CH_3NH_3PbI_3$ perovskite thin films determined by spectroscopic ellipsometry and spectrophotometry. *J. Phys. Chem. Lett.* **2014**, *6*, 66–71. [CrossRef]

39. Haynes, W.M. *CRC Handbook of Chemistry and Physics*; CRC Press: Boca Raton, FL, USA, 2014.
40. Ng, R.J.; Goh, X.M.; Yang, J.K. All-metal nanostructured substrates as subtractive color reflectors with near-perfect absorptance. *Opt. Express* **2015**, *23*, 32597–32605. [CrossRef]
41. Würfel, P.; Würfel, U. *Physics of Solar Cells: From Basic Principles to Advanced Concepts*; John Wiley & Sons: Hoboken, NJ, USA, 2016.

Article

Titanium Dioxide-Coated Zinc Oxide Nanorods as an Efficient Photoelectrode in Dye-Sensitized Solar Cells

Qiang Zhang [1], Shengwen Hou [2] and Chaoyang Li [1,2,*]

[1] School of Systems Engineering, Kochi University of Technology, Kami, Kochi 782-8502, Japan; 216005z@gs.kochi-tech.ac.jp

[2] Center for Nanotechnology, Kochi University of Technology, Kami, Kochi 782-8502, Japan; 186003p@gs.kochi-tech.ac.jp

* Correspondence: li.chaoyang@kochi-tech.ac.jp; Tel.: +81-887-57-2106

Received: 14 July 2020; Accepted: 12 August 2020; Published: 14 August 2020

Abstract: Well-arrayed zinc oxide nanorods applied as photoelectrodes for dye-sensitized solar cells were synthesized on an aluminum-doped zinc oxide substrate by the multi-annealing method. In order to improve the chemical stability and surface-to-volume ratio of photoanodes in dye-sensitized solar cells, the synthesized zinc oxide nanorods were coated with pure anatase phase titanium dioxide film using a novel mist chemical vapor deposition method. The effects of the titanium dioxide film on the morphological, structural, optical, and photovoltaic properties of zinc oxide–titanium dioxide core–shell nanorods were investigated. It was found that the diameter and surface-to-volume ratio of zinc oxide nanorods were significantly increased by coating them with titanium dioxide thin film. The power conversion efficiency of dye-sensitized solar cells was improved from 1.31% to 2.68% by coating titanium dioxide film onto the surface of zinc oxide nanorods.

Keywords: titanium dioxide; zinc oxide; core–shell nanorods; mist chemical vapor deposition; dye-sensitized solar cell

---

1. Introduction

Since Grätzel et al. developed the titanium dioxide (TiO_2)-based dye-sensitized solar cell (DSSC) in 1991 [1], the DSSC has emerged as a promising photovoltaic device, due to its promising power conversion efficiency (PCE), low fabrication cost, and low toxicity [2–5]. Hitherto, it has been reported that TiO_2-based DSSCs achieved a notable PCE of over 14% [6]. However, further improvements in PCE are difficult to achieve due to some disadvantages in current TiO_2-based DSSCs, such as the low carrier transportation rate of TiO_2 resulting from its low electron mobility, as well as the difficulty in fabricating TiO_2 nanostructures with a large surface-to-volume ratio [7,8]. Recently, zinc oxide (ZnO) has been widely investigated in different types of solar cells [9–11]. As an alternative photoanode material of DSSCs, ZnO has attracted much attention because it exhibits a similar bandgap and electron injection process from excited dye molecules to TiO_2 [12,13]. Moreover, the electron mobility of ZnO (200~1000 $cm^2/(V{\cdot}s)$) is much higher than that of TiO_2 (0.1~4 $cm^2/(V{\cdot}s)$) [14], which will enhance electron transportation. Additionally, compared with TiO_2, it is much easier to fabricate ZnO as various nanostructures to enlarge the surface-to-volume ratio [15]. Therefore, ZnO-based nanostructures and nanocomposites have much potential for application as a photoanode material to improve the PCE of DSSC.

However, the poor chemical stability of ZnO in the acidic dye solution and electrolyte solution of DSSCs has hampered its wider applicability as a photoanode material in DSSCs [16]. Additionally, defects easily form in ZnO, which increases the Zn^{2+}/dye complex and the electron–hole recombination at the interface [17–20]. In order to overcome the shortcomings of ZnO-based photoanodes, one solution is to coat a chemically stable shell onto the surface of as-deposited ZnO. This core–shell structure can passivate ZnO's surface to reduce the complex and form an energy barrier, thereby reducing the electron–hole recombination [21]. Among different ZnO-based nanocomposites, one of the most promising structures is ZnO–TiO_2's core–shell nanostructure. According to the literature [22–25], the PCE of ZnO photoanode-based DSSCs can be improved by about one to five times by replacing the ZnO photoanode with a corresponding ZnO–TiO_2 core–shell nanostructure. It is reported that ZnO's nanostructure could be coated with TiO_2 thin film using the sol–gel method [26], solution method [27], and atomic layer deposition [28]. However, the difficulties that arise with the uniformity and also in controlling the thickness of the TiO_2 layer are still unsolved.

Based on our previous study, DSSCs with a high PCE could be achieved by controlling the vertical alignment of ZnO nanorods and the quality of transparent conductive substrates [29–31]. In addition, mist chemical vapor deposition (mist CVD) has been proven to be an effective method for modifying ZnO nanorods [32,33]. In this study, ZnO nanorods with vertical alignment were fabricated by a multi-annealing process in reducing ambient. Compared with ZnO nanorods fabricated by other methods, the ZnO nanorods fabricated by multi-annealing showed a higher concentration of oxygen vacancies. The oxygen vacancies were generated due to the effect of reducing ambient and they enhanced the conductivity of ZnO nanorods. However, the oxygen vacancies on the surface of ZnO nanorods will trigger the recombination of electrons. In order to solve this issue, the TiO_2 thin layer was coated on ZnO nanorods by the mist CVD method to prevent the recombination of electrons and enhance the chemical stability of electrodes. Compared with other methods, the combination of the multiple annealing process and mist CVD method is an effective method to fabricate ZnO–TiO_2 core–shell nanorods applied as photoelectrodes for DSSCs. Figure 1 shows the fabrication mechanism and working principle of ZnO–TiO_2 core–shell nanorods. The electrons are injected from excited dye molecules to the conduction band (CB) of TiO_2. Then, the electrons are transferred from the CB of TiO_2 to the CB of ZnO. The ZnO core has high electron mobility and the TiO_2 shell can protect the ZnO core from corrosion and suppress the recombination of electrons. After coating, the obtained ZnO–TiO_2 core–shell nanorods, as well as the as-deposited ZnO nanorods, were used to fabricate DSSCs for comparison. The effects of TiO_2 coating on the properties of ZnO–TiO_2 core–shell nanorods were investigated in detail.

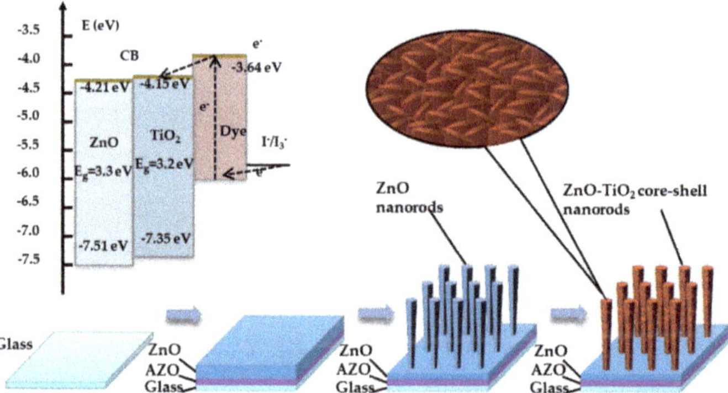

Figure 1. Fabrication mechanism and working principle of the zinc oxide–titanium dioxide (ZnO–TiO_2) core–shell nanorod.

2. Materials and Methods

2.1. Deposition of Thin Films

The aluminum-doped ZnO (AZO, 300 nm) thin films were deposited on alkali-free glass sheets (Eagle XG, Corning Inc., Corning, NY, USA) using a conventional radio frequency (RF, 13.56 MHz) magnetron sputtering system with an AZO target (2 wt.% Al_2O_3). Following the deposition of AZO films, ZnO films with a 500 nm thickness were deposited on AZO by the same sputtering system with a ZnO target (5N). Table 1 shows the deposition conditions of the AZO film and ZnO film. Argon was selected as the working gas, the flow rate of which was maintained at 30 sccm. During the deposition, the working distance and temperature were set and maintained at 60 mm and 150 °C, respectively. The pressure and RF power for AZO film deposition were maintained at 1 Pa and 60 W. For the deposition of ZnO, the pressure and RF power were held at 7 Pa and 180 W.

Table 1. Deposition conditions of AZO and ZnO films.

Film	AZO	ZnO
Target	AZO (2 wt.%)	ZnO (5N)
Working gas, Flow Rate (sccm)	Argon, 30	Argon, 30
Working distance (mm)	60	60
Deposition Temperature (°C)	150	150
Pressure (Pa)	1	7
RF Power (W)	60	180

2.2. *Fabrication of ZnO Nanorods*

After sputtering deposition, the fabricated ZnO films were treated using a multi-annealing process in a conventional annealing furnace. As shown in Table 2, the temperature was firstly kept at 300 °C for 2 h in a forming gas ambient (H_2:N_2 = 1.96%) to increase the density of zinc seeds on the surface. Then, the temperature was increased to 450 °C and kept at this level for 3 h for forming gas to produce the ZnO nanorods. Before the third forming gas annealing process, oxygen was introduced into the furnace for 40 min for surface oxidation to avoid an excessive reducing reaction. For safety considerations, nitrogen was introduced for 5 min between the forming gas and oxygen annealing processes.

Table 2. Annealing condition.

Step	Gas	Temperature (°C)	Time (min)
1	H_2 in N_2 (1.96%)	300	120
2	H_2 in N_2 (1.96%)	450	180
3	O_2	450	40
4	H_2 in N_2 (1.96%)	450	120

2.3. *Fabrication of ZnO–TiO_2 Core–Shell Nanorods*

Finally, TiO_2 film was coated onto the surface of the fabricated ZnO nanorods by a mist CVD system. Table 3 shows the deposition condition of the TiO_2 film. An ethanolic titanium tetraisopropoxide (TTIP, purity > 95.0%, Wako Pure Chemical Industries, Ltd., Osaka, Japan) solution with a concentration of 0.10 mol/L was prepared as the precursor solution. Mist droplets were generated from the precursor solution by ultrasonic atomization (2.4 MHz) and transferred to the reaction chamber by compressed air. The sample of as-deposited ZnO nanorods was placed in the reaction chamber and heated to 450 °C during the coating process.

Table 3. Deposition condition of TiO_2.

Solvent	Ethanol
Solute	TTIP
Concentration (mol/L)	0.10
Deposition Temperature (°C)	450
Carrier Gas, Flow Rate (L/min)	Compressed Air (dried), 2.5
Dilution Gas, Flow Rate (L/min)	Compressed Air (dried), 4.5

2.4. Fabrication of DSSC

The obtained ZnO–TiO_2 core–shell nanorods, as well as the as-deposited ZnO nanorods, were applied as photoanodes to fabricate DSSCs for comparison. N719 (Sigma Aldrich, St. Louis, MO, USA) was used as a dye sensitizer. The photoanodes were immersed in an ethanoic dye solution with a concentration of 5×10^{-4} mol/L for 12 h. A solution containing 0.10 mol/L lithium iodine and 0.05 mol/L iodine was used as the electrolyte. A platinum-coated indium-doped tin oxide film on glass was applied as the counter-electrode. Six samples of DSSCs were fabricated and investigated to confirm their reproducibility.

2.5. Characterization

The morphological properties of the AZO film, as-deposited ZnO nanorods, and ZnO–TiO_2 core–shell nanorods were evaluated using field emission scanning electron microscopy (FE-SEM, JSM-7400F, JEOL, Tokyo, Japan) and transmission emission microscopy (TEM, JEM 2100F, JEOL, Tokyo, Japan). The structural properties of the AZO film were measured by X-ray diffraction (XRD, ATX-G, Rigaku, Tokyo, Japan). The structural properties of the as-deposited ZnO nanorods and ZnO–TiO_2 core–shell nanorods were investigated by grazing incidence X-ray diffraction (GIXRD, ATX-G, Rigaku, Tokyo, Japan). The optical properties of the as-deposited ZnO nanorods and ZnO–TiO_2 core–shell nanorods were obtained using a spectrophotometer (U-4100, Hitachi, Tokyo, Japan). The fabricated DSSCs were characterized using a solar simulator (PEC-L01, AM 1.5 G, 100 mW/cm^2, Peccell Technologies Inc., Yokohama, Japan) and a source meter (Keithley 2400, Keithley Instruments Inc., Solon, OH, USA). All of the measurements were carried out at room temperature.

3. Results

The XRD pattern of the AZO film is shown in Figure 2. It was found that only the (002) diffraction peak was observed in the XRD pattern, which indicated that the AZO films had highly (002) preferred orientation with a c-axis perpendicular to the substrates. The insert image in Figure 2 shows the FE-SEM top view image of the AZO film. It is confirmed that an AZO film with a uniform surface was obtained after deposition.

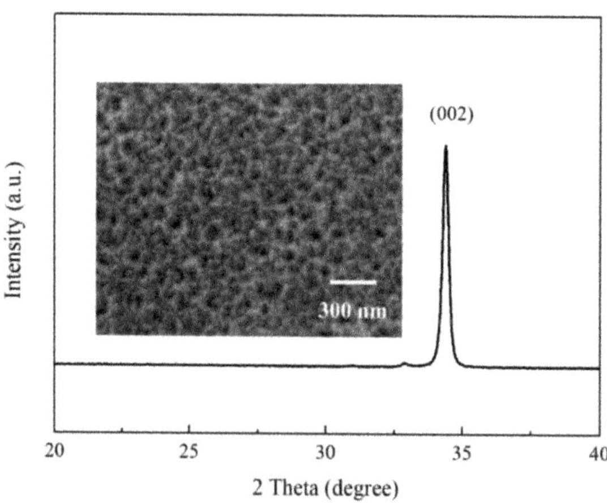

Figure 2. XRD pattern of AZO film (insert image shows the FE-SEM top view image of AZO film).

The FE-SEM images of the as-deposited ZnO nanorods and ZnO–TiO_2 core–shell nanorods are shown in Figure 3. The details of single nanorods are shown in the inset images. The as-deposited ZnO nanorods showed a well-arrayed hexagonal structure with a smooth surface. Compared with the as-deposited ZnO nanorods, the ZnO–TiO_2 core–shell nanorods had a higher surface roughness and a larger diameter. Intertwined TiO_2 nanosheets were observed on the surface of the ZnO–TiO_2 core–shell nanorods, indicating that the TiO_2 film was successfully coated onto the surface of the ZnO nanorods. Figure 3c shows the TEM image of a single ZnO–TiO_2 core–shell nanorod. It was confirmed that the thickness of the TiO_2 shell on the ZnO nanorods was around 15 nm.

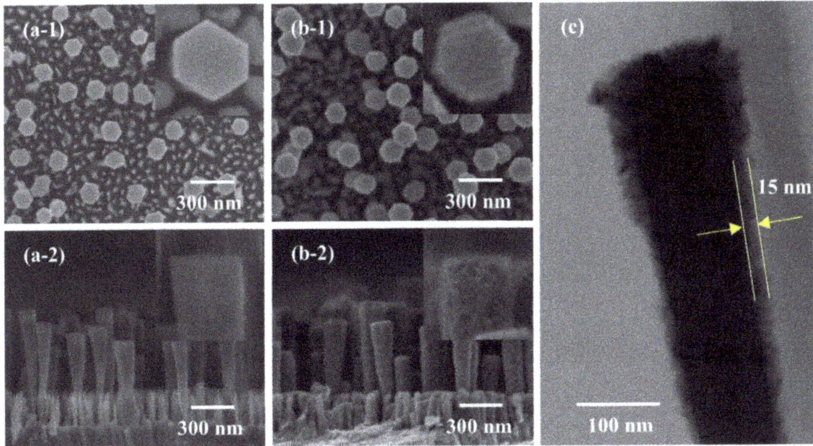

Figure 3. FE-SEM images of (**a**) as-deposited ZnO nanorods, (**b**) ZnO–TiO_2 core–shell nanorods, and (**c**) TEM image of single ZnO–TiO_2 core–shell nanorod ((**1**) Top view, (**2**) cross-section view).

The GIXRD patterns of the as-deposited ZnO nanorods and ZnO–TiO_2 core–shell nanorods are shown in Figure 4. It was found that only the (002) diffraction peak was observed in the GIXRD pattern of the as-deposited ZnO nanorods, suggesting that both the ZnO film and ZnO nanorods had highly (002) preferred orientation with a c-axis perpendicular to the substrates. This agrees well with the FE-SEM results. In the GIXRD pattern of the ZnO–TiO_2 core–shell nanorods, the observed peaks corresponded with the (101), (200), (211), (204), (220), and (215) diffraction peaks of the anatase phase TiO_2 and the (002) diffraction peak of ZnO. All of the diffraction peaks of TiO_2 were identified and corresponded with the anatase phase of TiO_2 (JCPDS 21-1272), indicating that the TiO_2 film coated on ZnO nanorods was pure anatase phase.

The optical transmission spectra of the as-deposited ZnO nanorods and ZnO–TiO_2 core–shell nanorods are shown in Figure 5. The as-deposited ZnO nanorods showed a high transmittance of 75% in visible range. After coating with TiO_2 film, the transmittance of the nanorods in visible range decreased to 55%, due to the scattering of TiO_2 nanosheets. It is well-known that the bandgap of material can be calculated from the transmittance data by the following equations [34,35]:

$$\alpha = \frac{1}{d}\ln\left(\frac{1}{T}\right) \tag{1}$$

$$(\alpha h\nu)^2 = A\left(h\nu - E_g\right) \tag{2}$$

where α is the absorption coefficient, d the thickness of material, T the transmittance, $h\nu$ the incident photon energy, A a constant, and E_g the bandgap. A plot of $(\alpha h\nu)^2$ as a function of $h\nu$ made to determine E_g by linear fitting is shown in Figure 6. After fitting, the bandgap of the as-deposited ZnO nanorods was determined as around 3.32 eV, corresponding with the bandgap of bulk ZnO (3.37 eV). The bandgap of the ZnO–TiO_2 core–shell nanorods was around 3.28 eV, corresponding with the bandgap of anatase phase TiO_2 (3.2 eV).

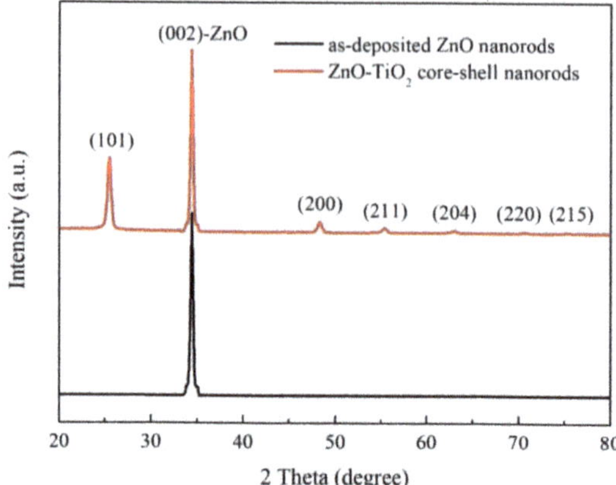

Figure 4. GIXRD patterns of as-deposited ZnO nanorods and ZnO–TiO_2 core–shell nanorods.

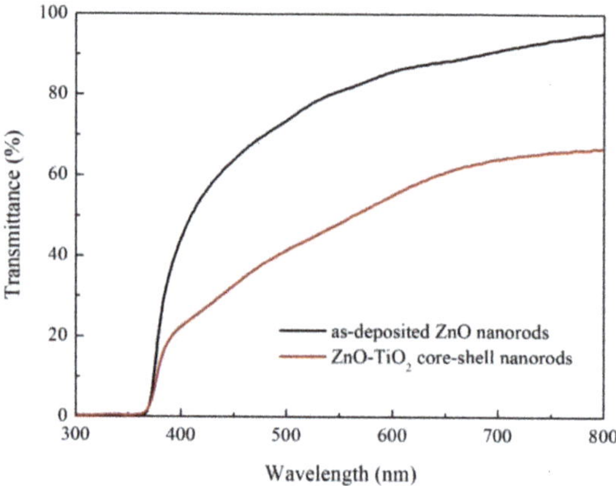

Figure 5. Optical transmission spectra of as-deposited ZnO nanorods and ZnO–TiO_2 core–shell nanorods.

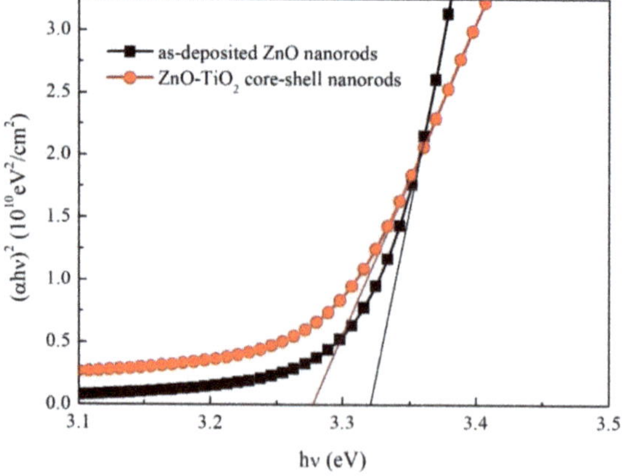

Figure 6. Variation of $(\alpha h\nu)^2$ of the as-deposited ZnO nanorods and ZnO–TiO_2 core–shell nanorods as a function of the photon energy $(h\nu)$.

Figure 7 shows the *J-V* characteristics of the demonstrated DSSCs applying as-deposited ZnO nanorods and ZnO–TiO_2 core–shell nanorods as photoanodes. Compared with the DSSCs using as-deposited ZnO nanorods, the DSSCs applying ZnO–TiO_2 core–shell nanorods showed higher open circuit voltage (V_{OC}), higher short circuit current density (J_{SC}), higher fill factor (*FF*), and higher PCE. After coating with TiO_2, the V_{OC} of the DSSCs increased from 0.60 V to 0.63 V, and the J_{SC} increased from 5.01 mA/cm^2 to 6.73 mA/cm^2. It was found that the *FF* increased from 43.41% to 63.13%, and the PCE increased from 1.31% to 2.68%. The results showed good reproducibility by checking all of the DSSCs samples. The significant improvement of the *FF* and PCE was due to the great improvement in the J_{SC}, which could be explained as follows: Firstly, the TiO_2 shell increased the surface-to-volume ratio of the ZnO nanorods. Therefore, more dye molecules were absorbed onto the surface of the nanorods, which enhanced their light harvesting. Secondly, the TiO_2 shell has a much lower electron–hole recombination rate than ZnO nanorods, which could greatly improve the efficiency of electron collection. Thirdly, the last step of a multi-annealing process was carried out in a reducing ambient. Consequently, many oxygen vacancies were generated on the surface of the ZnO nanorods. The oxygen vacancies acted as recombination centers, which triggered large amounts of recombination of electrons. After coating with TiO_2 film, the recombination of electrons was suppressed. The efficient light harvesting and efficient electron collection contributed to the great improvement in the J_{SC}.

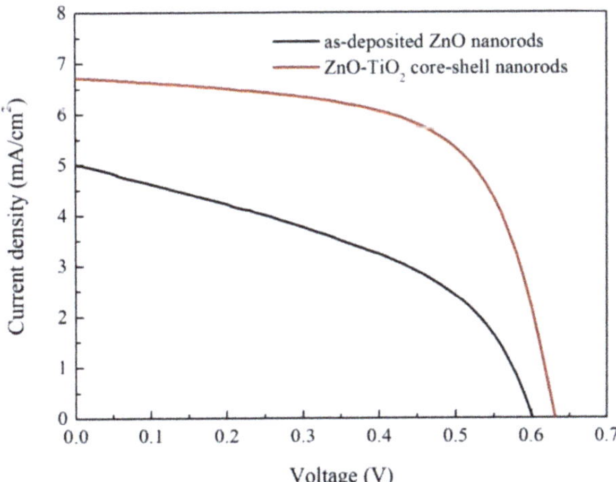

Figure 7. *J-V* characteristics of demonstrated dye-sensitized solar cells (DSSCs) applying as-deposited ZnO nanorods and ZnO–TiO_2 core–shell nanorods as photoanodes.

4. Conclusions

Well-arrayed ZnO–TiO_2 core–shell nanorods were successfully synthesized on AZO substrates by RF magnetron sputtering, multi-annealing, and the mist CVD method. The morphology of the ZnO nanorods was significantly changed by coating with a TiO_2 film. After forming the ZnO–TiO_2 core–shell structures, the diameter and surface-to-volume ratio of the nanorods were greatly increased. The PCE of DSSCs applying ZnO nanorods as photoanodes was increased two-fold from 1.31% to 2.68% by coating with TiO_2.

Author Contributions: Conceptualization, S.H. and C.L.; methodology, S.H. and C.L.; validation, S.H. and C.L.; investigation, S.H.; data curation, Q.Z. and S.H.; writing—original draft preparation, Q.Z.; writing—review and editing, Q.Z. and C.L.; visualization, Q.Z.; supervision, C.L.; project administration, C.L.; funding acquisition, C.L. All authors have read and agreed to the published version of the manuscript.

Funding: This research was funded by the Ministry of Education, Culture, Sports, Science, and Technology (MEXT) in Japan, grant number [17K06394].

Acknowledgments: The authors gratefully acknowledge the financial support by Grant-in-Aid for Scientific Research from the Ministry of Education, Culture, Sports, Science and Technology (MEXT), Japan. The authors greatly appreciate the assistance of Associate Professor Noriko Nitta in supporting the TEM measurement.

Conflicts of Interest: The authors declare no conflict of interest.

References

1. O'Regan, B.; Grätzel, M. A low-cost, high-efficiency solar cell based on dye-sensitized. *Nature* **1991**, *353*, 737–740. [CrossRef]
2. Prabavathy, N.; Shalini, S.; Balasundaraprabhu, R.; Velauthapillai, D.; Prasanna, S.; Muthukumarasamy, N. Enhancement in the photostability of natural dyes for dye-sensitized solar cell (DSSC) applications: A review. *Int. J. Energy Res.* **2017**, *41*, 1372–1396. [CrossRef]
3. Ahmad, M.S.; Pandey, A.K.; Rahim, N.A. Advancements in the development of TiO_2 photoanodes and its fabrication methods for dye sensitized solar cell (DSSC) applications. A review. *Renew. Sustain. Energy Rev.* **2017**, *77*, 89–108. [CrossRef]
4. Das, T.K.; Ilaiyaraja, P.; Sudakar, C. Template assisted nanoporous TiO_2 nanoparticles: The effect of oxygen vacancy defects on photovoltaic performance of DSSC and QDSSC. *Sol. Energy* **2018**, *159*, 920–929. [CrossRef]
5. Vaghasiya, J.V.; Sonigara, K.K.; Soni, S.S.; Tan, S.C. Dual functional hetero-anthracene based single component organic ionic conductors as redox mediator cum light harvester for solid state photoelectrochemical cells. *J. Mater. Chem. A* **2018**, *6*, 4868–4877. [CrossRef]
6. Kakiage, K.; Aoyama, Y.; Yano, T.; Oya, K.; Fujisawa, J.-I.; Hanaya, M. Highly-efficient dye-sensitized solar cells with collaborative sensitization by silyl-anchor and carboxy-anchor dyes. *Chem. Commun.* **2015**, *51*, 15894–15897. [CrossRef]
7. Law, M.; Greene, L.E.; Johnson, J.C.; Saykally, R.; Yang, P. Nanowire dye-sensitized solar cells. *Nat. Mater.* **2005**, *4*, 455–459. [CrossRef]
8. Zhang, Q.; Dandeneau, C.S.; Zhou, X.; Cao, G. ZnO nanostructures for dye-sensitized solar cells. *Adv. Mater.* **2009**, *21*, 4087–4108. [CrossRef]
9. Gopalan, S.-A.; Gopalan, A.-I.; Vinu, A.; Lee, K.-P.; Kang, S.-W. Solar Energy Materials and Solar Cells A new optical-electrical integrated bu ff er layer design based on gold nanoparticles tethered thiol containing sulfonated polyaniline towards enhancement of solar cell performance. *Sol. Energy Mater. Sol. Cells* **2018**, *174*, 112–123. [CrossRef]
10. Nandakumar, D.K.; Vaghasiya, J.V.; Yang, L.; Zhang, Y.; Tan, S.C. A solar cell that breathes in moisture for energy generation. *Nano Energy* **2020**, *68*, 104263. [CrossRef]
11. Nandakumar, D.K.; Ravi, S.K.; Zhang, Y.; Guo, N.; Zhang, C.; Tan, S.C. A super hygroscopic hydrogel for harnessing ambient humidity for energy conservation and harvesting. *Energy Environ. Sci.* **2018**, *11*, 2179–2187. [CrossRef]
12. Kolodziejczak-Radzimska, A.; Jesionowski, T. Zinc oxide-from synthesis to application: A review. *Materials (Basel)* **2014**, *7*, 2833–2881. [CrossRef]
13. Lee, J.-C.; Gopalan, A.-I.; Saianand, G.; Lee, K.-P.; Kim, W.-J. Manganese and graphene included titanium dioxide composite nanowires: Fabrication, characterization and enhanced photocatalytic activities. *Nanomaterials* **2020**, *10*, 456. [CrossRef]
14. Tiwana, P.; Docampo, P.; Johnston, M.B.; Snaith, H.J.; Herz, L.M. Electron mobility and injection dynamics in mesoporous ZnO, SnO_2, and TiO_2 films used in dye-sensitized solar cells. *ACS Nano.* **2011**, *5*, 5158–5166. [CrossRef]
15. Gonzalez-Valls, I.; Lira-Cantu, M. Vertically-aligned nanostructures of ZnO for excitonic solar cells: A review. *Energy Environ. Sci.* **2009**, *2*, 19–34. [CrossRef]
16. Vittal, R.; Ho, K.-C. Zinc oxide based dye-sensitized solar cells: A review. *Renew. Sustain. Energy Rev.* **2017**, *70*, 920–935. [CrossRef]
17. Lu, L.; Li, R.; Fan, K.; Peng, T. Effects of annealing conditions on the photoelectrochemical properties of dye-sensitized solar cells made with ZnO nanoparticles. *Sol. Energy.* **2010**, *84*, 844–853. [CrossRef]
18. Ambade, S.B.; Mane, R.S.; Ghule, A.V.; Takwale, M.G.; Abhyankar, A.; Cho, B.W.; Han, S.H. Contact angle measurement: A preliminary diagnostic method for evaluating the performance of ZnO platelet-based dye-sensitized solar cells. *Scr. Mater.* **2009**, *61*, 12–15. [CrossRef]

19. Yan, F.; Huang, L.; Zheng, J.; Huang, J.; Lin, Z.; Huang, F.; Wei, M. Effect of surface etching on the efficiency of ZnO-based dye-sensitized solar cells. *Langmuir.* **2010**, *26*, 7153–7156. [CrossRef]
20. Horiuchi, H.; Katoh, R.; Hara, K.; Yanagida, M.; Murata, S.; Arakawa, H.; Tachiya, M. Electron injection efficiency from excited N3 into nanocrystalline ZnO films: Effect of (N3-Zn^{2+}) aggregate formation. *J. Phys. Chem. B.* **2003**, *107*, 2570–2574. [CrossRef]
21. Law, M.; Greene, L.E.; Radenovic, A.; Kuykendall, T.; Liphardt, J.; Yang, P. ZnO-Al_2O_3 and ZnO-TiO_2 core-shell nanowire dye-sensitized solar cells. *J. Phys. Chem. B.* **2006**, *110*, 22652–22663. [CrossRef] [PubMed]
22. Chandiran, A.K.; Abdi-Jalebi, M.; Nazeeruddin, M.K.; Grätzel, M. Analysis of electron transfer properties of ZnO and TiO_2 photoanodes for dye-sensitized solar cells. *ACS Nano.* **2014**, *8*, 2261–2268. [CrossRef] [PubMed]
23. Atienzar, P.; Ishwara, T.; Illy, B.N.; Ryan, M.P.; O'Regan, B.C.; Durrant, J.R.; Nelson, J. Control of photocurrent generation in polymer/ZnO nanorod solar cells by using a solution-processed TiO_2 overlayer. *J. Phys. Chem. Lett.* **2010**, *1*, 708–713. [CrossRef]
24. Feng, Y.; Ji, X.; Duan, J.; Zhu, J.; Jiang, J.; Ding, H.; Meng, G.; Ding, R.; Liu, J.; Hu, A.; et al. Synthesis of ZnO@TiO_2 core-shell long nanowire arrays and their application on dye-sensitized solar cells. *J. Solid State Chem.* **2012**, *190*, 303–308. [CrossRef]
25. Prabakar, K.; Son, M.; Kim, W.-Y.; Kim, H. TiO_2 thin film encapsulated ZnO nanorod and nanoflower dye sensitized solar cells. *Mater. Chem. Phys.* **2011**, *125*, 12–14. [CrossRef]
26. Zhao, R.; Zhu, L.; Cai, F.; Yang, Z.; Gu, X.; Huang, J.; Cao, L. ZnO/TiO_2 core-shell nanowire arrays for enhanced dye-sensitized solar cell efficiency. *Appl. Phys. A Mater. Sci. Process.* **2013**, *113*, 67–73. [CrossRef]
27. Goh, G.K.L.; Le, H.Q.; Huang, T.J.; Hui, B.T.T. Low temperature grown ZnO@TiO_2 core shell nanorod arrays for dye sensitized solar cell application. *J. Solid State Chem.* **2014**, *214*, 17–23. [CrossRef]
28. Greene, L.E.; Law, M.; Yuhas, B.D.; Yang, P. ZnO-TiO_2 core-shell nanorod/P3HT solar cells. *J. Phys. Chem. C.* **2007**, *111*, 18451–18456. [CrossRef]
29. Li, X.; Li, C.; Kawaharamura, T.; Wang, D.; Nitta, N.; Furuta, M.; Furuta, H.; Hatta, A. Influence of substrates on formation of zinc oxide nanostructures by a novel reducing annealing method. *Nanosci. Nanotechnol. Lett.* **2014**, *6*, 174–180. [CrossRef]
30. Li, X.; Li, C.; Hou, S.; Hatta, A.; Yu, J.; Jiang, N. Thickness of ITO thin film influences on fabricating ZnO nanorods applying for dye-sensitized solar cell. *Compos. Part B Eng.* **2015**, *74*, 147–152. [CrossRef]
31. Hou, S.; Li, C. Aluminum-doped zinc oxide thin film as seeds layer effects on the alignment of zinc oxide nanorods synthesized in the chemical bath deposition. *Thin Solid Films* **2016**, *605*, 37–43. [CrossRef]
32. Li, X.; Li, C.; Kawaharamura, T.; Wang, D.; Nitta, N.; Furuta, M.; Furuta, H.; Hatta, A. Fabrication of zinc oxide nanostructures by mist chemical vapor deposition. *Trans. Mater. Res. Soc. Jpn.* **2014**, *164*, 161–164. [CrossRef]
33. Zhang, Q.; Li, C. TiO_2 coated ZnO nanorods by mist chemical vapor deposition for application as photoanodes for dye-sensitized solar cells. *Nanomaterials* **2019**, *9*, 1339. [CrossRef] [PubMed]
34. Gao, X.; Du, Y.; Meng, X. Cupric oxide film with a record hole mobility of 48.44 cm^2/Vs via direct-current reactive magnetron sputtering for perovskite solar cell application. *Sol. Energy* **2019**, *191*, 205–209. [CrossRef]
35. Fang, X.S.; Bando, Y.; Shen, G.Z.; Ye, C.H.; Gautam, U.K.; Costa, P.M.F.J.; Zhi, C.Y.; Tang, C.C.; Golberg, D. Ultrafine ZnS nanobelts as field emitters. *Adv. Mater.* **2007**, *19*, 2593–2596. [CrossRef]

 nanomaterials

MDPI

Article

Temperature Dependence of Carrier Extraction Processes in GaSb/AlGaAs Quantum Nanostructure Intermediate-Band Solar Cells

Yasushi Shoji [1,2,*], Ryo Tamaki [2] and Yoshitaka Okada [2]

[1] Global Zero Emission Research Center, National Institute of Advanced Industrial Science and Technology (AIST), Tsukuba, Ibaraki 305-8568, Japan

[2] Research Center for Advanced Science and Technology (RCAST), The University of Tokyo, Meguro-ku, Tokyo 153-8904, Japan; tamaki@mbe.rcast.u-tokyo.ac.jp (R.T.); okada@mbe.rcast.u-tokyo.ac.jp (Y.O.)

* Correspondence: y.shoji@aist.go.jp; Tel.: +81-29-861-8251

Abstract: From the viewpoint of band engineering, the use of GaSb quantum nanostructures is expected to lead to highly efficient intermediate-band solar cells (IBSCs). In IBSCs, current generation via two-step optical excitations through the intermediate band is the key to the operating principle. This mechanism requires the formation of a strong quantum confinement structure. Therefore, we focused on the material system with GaSb quantum nanostructures embedded in AlGaAs layers. However, studies involving crystal growth of GaSb quantum nanostructures on AlGaAs layers have rarely been reported. In our work, we fabricated GaSb quantum dots (QDs) and quantum rings (QRs) on AlGaAs layers via molecular-beam epitaxy. Using the Stranski–Krastanov growth mode, we demonstrated that lens-shaped GaSb QDs can be fabricated on AlGaAs layers. In addition, atomic force microscopy measurements revealed that GaSb QDs could be changed to QRs under irradiation with an As molecular beam even when they were deposited onto AlGaAs layers. We also investigated the suitability of GaSb/AlGaAs QDSCs and QRSCs for use in IBSCs by evaluating the temperature characteristics of their external quantum efficiency. For the GaSb/AlGaAs material system, the QDSC was found to have slightly better two-step optical excitation temperature characteristics than the QRSC.

Keywords: intermediate-band solar cell; quantum dot; molecular-beam epitaxy

Citation: Shoji, Y.; Tamaki, R.; Okada, Y. Temperature Dependence of Carrier Extraction Processes in GaSb/AlGaAs Quantum Nanostructure Intermediate-Band Solar Cells. *Nanomaterials* **2021**, *11*, 344. https://doi.org/10.3390/nano11020344

Academic Editor: Efrat Lifshitz

Received: 28 December 2020
Accepted: 25 January 2021
Published: 29 January 2021

1. Introduction

With increasing interest in a low-carbon society, the development of solar cells (SCs) with high power generation efficiency has become urgent. As novel photovoltaic devices for realizing a high conversion efficiency that exceeds the limit of conventional SCs, intermediate-band solar cells (IBSCs) have recently become the focus of attention [1–4]. The operating principle of IBSCs involves optical transition via sub-bandgap intermediate-band (IB), which can absorb photons with an energy that is less than the difference between the conduction band (CB) and valence band (VB) of the host material, thereby increasing the photocurrent. The IB formed between CB and VB has also been used to obtain interfacial conductivity in wide-gap material devices, in which IB functions as a pathway for charge transfer [5]. However, in IBSCs, two-step optical absorption through IB is key to the operation. The theoretical conversion efficiency of standard single-junction SCs without IB depends on the bandgap of the absorber. For SCs with small bandgaps, a high output current can be obtained because there are few photons transmitted through them, whereas the output voltage becomes small because it is determined by the quasi-Fermi level splitting of the absorber. The opposite occurs if the band gap of the absorber is large. In ideal IBSCs, the quasi-Fermi levels of CB, VB, and IB are independent of each other. This is achieved by the carrier transitions of VB–IB and IB–CB, which occur by optical excitations rather than thermal excitations. As a result, ideal IBSCs can obtain a current gain without reducing the host material voltage owing to the absorption of photons with energy lower than the bandgap of the host material using IB. The use of a quantum dot (QD) superlattice has been intensively investigated as a method for forming IBs [6–10]. Although photocurrent production by two-step optical excitation via QD states in QDSCs has been reported, the contribution of this two-step excitation to the conversion efficiency is still small [11–13].

Ideally, in the process of extracting carriers generated in the IB, the processes involving thermal excitation and electric-field assistance should be suppressed as much as possible and the processes involving optical excitation should be dominant. Therefore, the following structure is preferable for QD-IBSCs: First, to prevent electric-field-induced carrier escape from the IB, the application of the strong built-in electric field of the *p–n* junction to the IB should be avoided. We previously reported that the electric-field strength applied to the IB can be controlled by a field-damping layer with a low carrier concentration and that the IB can thus be formed nearly flat [14,15].

Second, a higher band offset energy between the QDs and barrier layers is needed to suppress the thermal excitation of carriers in the IB. This is important for obtaining the high-temperature operation of two-step optical excitation via IB. Since solar cells are usually used for terrestrial applications at room temperature or higher, the thermal escape rate of carriers absorbed within IB increases due to the high thermal energy at such temperatures. As mentioned above, the main principle of IBSC is to obtain the current gain without any voltage loss using optical transitions through IB, which can be achieved by separating the quasi-Fermi levels of CB, VB, and IB. A high thermal escape rate case means that the quasi-Fermi levels are not well separated, leading to a decrease in the output voltage [16]. InAs quantum dots embedded by GaAs layers are among the most typical QDs in III-V materials, and their growth technology is more mature than that of other materials. Therefore, IBSCs have been vigorously studied in InAs/GaAs QDs. However, effective current production due to two-step absorption has not been achieved at higher temperatures because of the small band offset between QD layers and the host material of GaAs. Ramiro et al. demonstrated the suppression of the thermal escape of carriers generated in QDs using wide-gap AlGaAs as a host material [17]. They reported that the higher band offset between InAs and AlGaAs suppresses the carrier thermal escape up to a temperature of 220 K. In addition, our previous work demonstrated the importance of increasing the band offset to obtain two-step optical absorption at higher temperatures [18], where the threshold temperature of the two-step optical absorption was defined as the temperature of the photocurrent intensity with 1% of the signal obtained at 9 K. According to this evaluation, changing the host material of InAs QDs from GaAs to wider-gap AlGaAs improves the threshold temperature of the two-step absorption from 82 to 139 K. Furthermore, a threshold temperature of 220 K was obtained in InGaAs/AlGaAs QDSCs, which were tuned for stronger confinement. If not limited to III-V materials, two-step optical absorption at room temperature has recently been obtained in IBSCs with lead sulfide QDs and wide-gap lead halide perovskites [19]. However, perovskite materials have some issues in terms of long-term stability. Among III-V semiconductors, wide-gap materials include AlGaAs, InGaP, and GaAsP. When phosphide materials are used as barrier layers for InAs QDs, the interface quality is often a problem due to the intermixing of As and P atoms. Therefore, a method for introducing a thin interlayer to improve the interface quality between QDs and barrier layers has been proposed [20–22].

Third, the carriers in the IB should have a sufficiently long lifetime to avoid rapid recombination. Therefore, a type-II quantum structure, in which electrons and holes are spatially separated, is preferable [23–27]. GaSb/AlGaAs type-II QDs are expected to exhibit the aforementioned preferred features. In this system, the holes are confined in GaSb QDs surrounded by AlGaAs barrier layers [28]. More interestingly, GaSb QDs can undergo a change in shape to quantum rings (QRs) when irradiated with an As molecular beam because of the desorption of Sb from the QDs, which occurs via As/Sb exchange [29,30]. We previously reported that GaSb QRs have less local lattice strain than GaSb QDs, resulting in better overall crystal quality of the multi stacked structure in a GaSb/GaAs system [31]. In the present paper, we first investigate the growth and structural properties of GaSb quantum nanostructures on an AlGaAs layer and then discuss whether the AlGaAs barrier layer is effective in improving two-step optical excitation in the GaSb QDs. In addition, the difference in the temperature dependence of the two-step optical excitation between GaSb/AlGaAs QDs and QRs is investigated.

2. Experimental Procedure

All samples were grown on GaAs (001) substrates via solid-source molecular-beam epitaxy. We first attempted to grow GaSb quantum nanostructures on an $Al_{0.2}Ga_{0.8}As$ layer. After the semi-insulating GaAs substrate was thermally cleaned at 590 °C, a 250-nm-thick buffer layer was grown at 580 °C. Then, 2.5 monolayers (MLs) GaSb was grown on a 200-nm-thick $Al_{0.2}Ga_{0.8}As$ layer at 480 °C. The Sb beam pressure was 6×10^{-5} Pa. The growth rate of GaSb was 0.6 ML/s. We confirmed that self-organized GaSb QDs could be formed via the Stranski–Krastanov growth mode under these growth conditions [32,33]. Moreover, samples based on the aforementioned structure were prepared by irradiation with an As molecular beam with a beam pressure 1×10^{-4} Pa for 0–30 s after GaSb growth to investigate the shape change of the quantum nanostructures to QRs. Similarly, samples with GaSb nanostructures grown at a substrate temperature of 460 °C were also prepared.

We also fabricated *p–i–n*-structured QD- and QR-IBSCs using AlGaAs as a host material. Figure 1 shows the device design of the IBSCs. After the *n*-GaAs substrate was thermally cleaned at 590 °C, a 250-nm-thick *n*-GaAs buffer layer, a 100-nm-thick *n*-$Al_{0.6}Ga_{0.4}As$ back-surface field (BSF) layer, and a 500-nm-thick *n*-$Al_{0.2}Ga_{0.8}As$ base layer were grown at 580 °C. The $Al_{0.6}Ga_{0.4}As$ BSF layer was introduced to reduce the recombination of minority carriers near the backside. Then, 10 layers of GaSb/$Al_{0.2}Ga_{0.8}As$ QDs or QRs were subsequently introduced into the *i*-layer. The multi stacked structures were grown at 460 °C. The 150-nm-thick *p*-$Al_{0.2}Ga_{0.8}As$ emitter layer, 30-nm-thick *p*-$Al_{0.8}Ga_{0.2}As$ window layer, and a 50-nm-thick *p*-GaAs contact layer were then grown. In all of the layers, Be and Si were used as *p*-type and *n*-type dopants, respectively. Ga and Al were evaporated using hot-lip-type dual-filament and single-filament effusion cells, respectively. As and Sb were supplied by cracker effusion cells that enabled valve control of the flux.

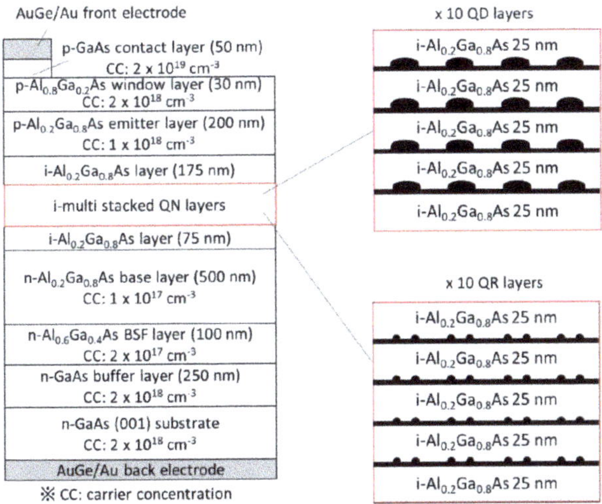

Figure 1. Schematic of the cell structures fabricated in the present study. The device structures consist of *p–i–n* layers. Multilayers of GaSb/AlGaAs quantum nanostructures were inserted into the *i*-layer.

After the growth, AuGe/Au was used as *n*-type and *p*-type Ohmic electrodes. The mesa-type structures were formed by chemical wet-etching using a solution with a H_3PO_4:H_2O_2:H_2O ratio of 4:1:20. Finally, a *p*-type GaAs contact layer was etched by a citric-acid-based solution in areas where the electrodes were not deposited.The structural properties of the quantum nanostructures were analyzed by atomic force microscopy (AFM) (Seiko Instruments Inc., E-Sweep, Chiba, Japan). The device performance was characterized using current–voltage curves measured under air mass 1.5 global (AM 1.5G) solar spectrum illumination. The optical characteristics were evaluated using a unique external quantum efficiency (EQE) measurement system (Figure 2). One of the light sources provided continuous-wave monochromatic photons that generated electron–hole pairs in the host material and quantum nanostructures. The EQE spectra were acquired by changing the wavelength of this monochromatic light. The other light source was a chopped infrared (IR) lamp with a long-pass filter that allowed only photons with a wavelength λ longer than 1.3 μm to pass through. The illumination of low-energy photons from this IR source provided the broadband excitation and pumped holes from the IB formed by quantum nanostructures to the VB. The photocurrent production due to two-step optical excitation via the IB was estimated by measuring the difference between the EQE with and without this IR illumination. The difference in EQE (ΔEQE) under IR illumination was detected using a lock-in amplifier synchronized (NF Corp. LI5655, Yokohama, Japan) with an optical chopper (Thorlabs Inc. MC2000, Newton, NJ, USA) set to 85 Hz. The measurement was performed in a cryostat.

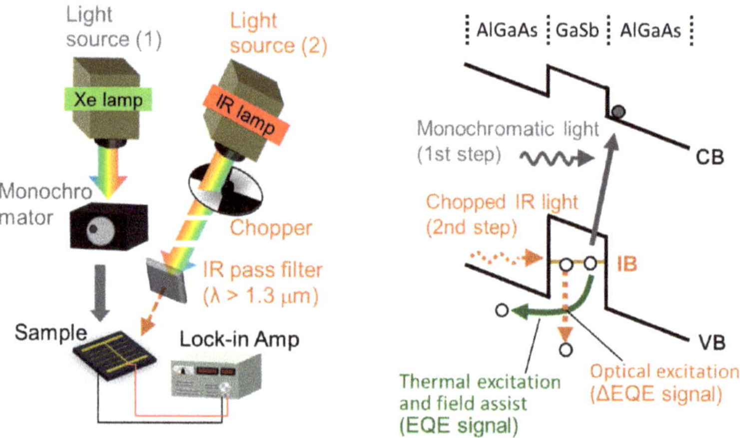

Figure 2. Measurement setup for characterization of photocurrent production due to each carrier extraction process. One of the light sources provides continuous-wave monochromatic photons that generate electron–hole pairs in the VB–CB and IB–CB (Light source (1)). Chopped IR lamp with a long-pass filter positioned at the exit passes through a set of filters that collectively allow only photons of $\lambda > 1300$ nm to be transmitted (Light source (2)). The illumination of low-energy photons from this IR source can then only pump the electrons from the IB to the VB.

3. Results and Discussion

3.1. Structural Properties of GaSb Quantum Nanostructures Grown on AlGaAs Layers

We first conducted AFM measurements to confirm the structure of the GaSb QDs and QRs grown on $Al_{0.2}Ga_{0.8}As$ layers. Figure 3 shows an AFM image for a single layer of GaSb quantum nanostructures grown on an $Al_{0.2}Ga_{0.8}As$ layer at 480 °C. These samples were fabricated at different As-soaking times after growth of the GaSb. Lens-shaped QD structures were observed for the sample without As-soaking, whereas a mixture of QDs, QRs, and their intermediate structures was observed for the sample with an As-soaking time of 10 s. For the samples with an As-soaking time of 22 and 30 s after growth of the

GaSb QDs, all of the QDs were transformed into QRs via the As/Sb exchange reaction. This result suggests that the shape of the GaSb quantum nanostructures can be changed by the As-soaking, even when the nanostructures are on an $Al_{0.2}Ga_{0.8}As$ layer. The exchange of the As and Sb atoms occurred because the bond between Ga and Sb atoms was much weaker than the bond between Ga and As atoms [34]. Lin et al. reported the transformation mechanism of GaSb QDs to QRs [30]. In their study, Sb with background As soaking was performed after the growth of GaSb QDs. During the soaking, As atoms were aggregated on the summit of the GaSb QDs, and such As-rich region induced lattice mismatch strain to the underlying GaSb QD. Hence, the Sb atoms were diffused away to relax the strain energy. Due to the larger strain on the summits of the QDs, most of the Sb atoms were repelled from the QDs. As a result, a ring structure was formed. The detailed structure of GaSb QR was reported by Timm et al., who analyzed the composition of QR and observed a clear central opening structure through scanning tunneling microscopy (STM) [29].

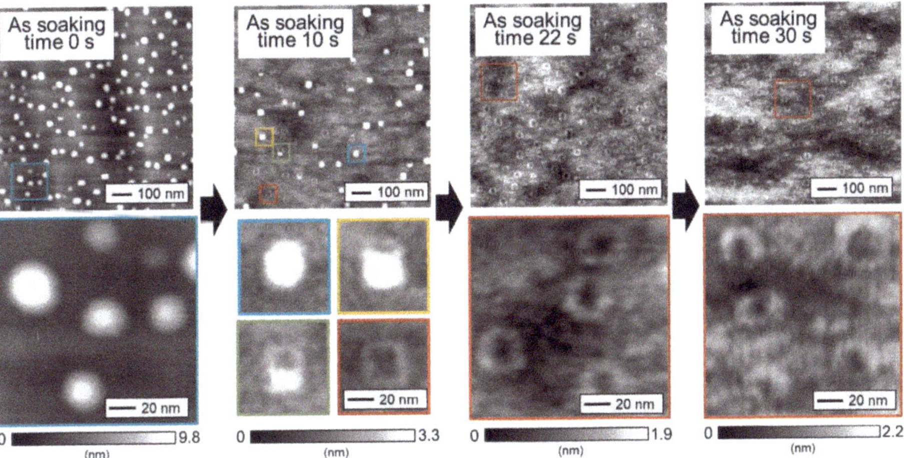

Figure 3. Plane view of AFM images for GaSb quantum nanostructures grown on an AlGaAs layer at 480 °C. The As-soaking time after the deposition of GaSb was varied.

In the present work, the transition from QDs to QRs appears to proceed as shown in Figure 4 because quantum nanostructures of various shapes are mixed together in the samples with an As-soaking time of 10 s. Immediately after the supply of As after the growth of GaSb QDs, the QD height decreases. The transition then progresses to a shape change as the height falls below a certain threshold. Subsequently, according to the transformation mechanism described above, the structure stabilizes in the ring shape when sufficient As atoms have been supplied. In the sample without As-soaking, QDs of various heights were formed, with an average height of 5.70 nm (in-plane density: 1.7×10^{10} cm^{-2}). Therefore, the extent of the shape change differed for each QD size. In the case of an As-soaking time of 10 s, QDs with a large height decreased in height but did not undergo a change in shape. However, the QDs with a small height decreased sufficiently in height for their shape to change even with a short As irradiation time of 10 s, leading to the formation of a ring shape. For the sample with an As-soaking time of 22 s, all of the QDs in the plane transformed into QRs because sufficient As was supplied to induce even the larger QDs to transform into QRs. The AFM image of the sample with the As-soaking time of 10 s shows the change process to QR. As shown in the shape change process of Figure 4, unlike the described trends in previous reports, a ring shape was formed from the side of the QDs rather than from the center, considerably due to the fact that the lattices of the GaSb crystal were anisotropically strained. Actually, the QDs grown in the present study did not have an isotropic shape, and the shape change by the As-soaking proceeded from the longitudinal direction. However, to clarify the main cause and prove the hypothesis, a further detailed structural analysis, such as STM, would be necessary.

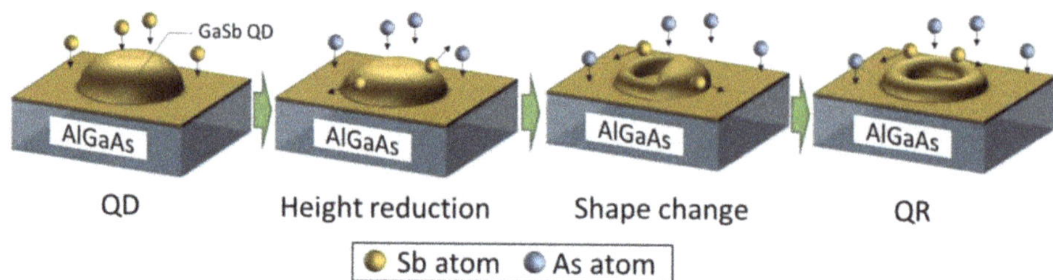

Figure 4. A model for the transformation of GaSb QDs into QRs, as inferred from this study.

Because growing multi stacked structures with large QDs for IBSC is difficult, we next attempted to grow the small QDs with a high in-plane density by decreasing the growth temperature. In addition, we verified that QR structures can also be fabricated at high density. Figure 5a shows GaSb quantum nanostructures grown on an AlGaAs layer at 460 °C. Table 1 shows the mean diameter, height, and density of QDs and QRs, as determined from AFM measurements. Herein, for the QRs, the outer diameters were measured. For the sample grown at 460 °C, small QDs were confirmed to form at a density greater than in the sample grown at 480 °C. This greater density is attributed to the lower growth temperature shortening the surface migration length of Ga atoms and facilitating the formation of small QDs. Moreover, the density of GaSb QRs also increased with decreasing growth temperature. Figure 5b shows the line profiles of a single QD and QR. The diameter of the QR structure was approximately the same as that of the QD structure, whereas the height of the QR structure was substantially smaller than that of the QD structure.

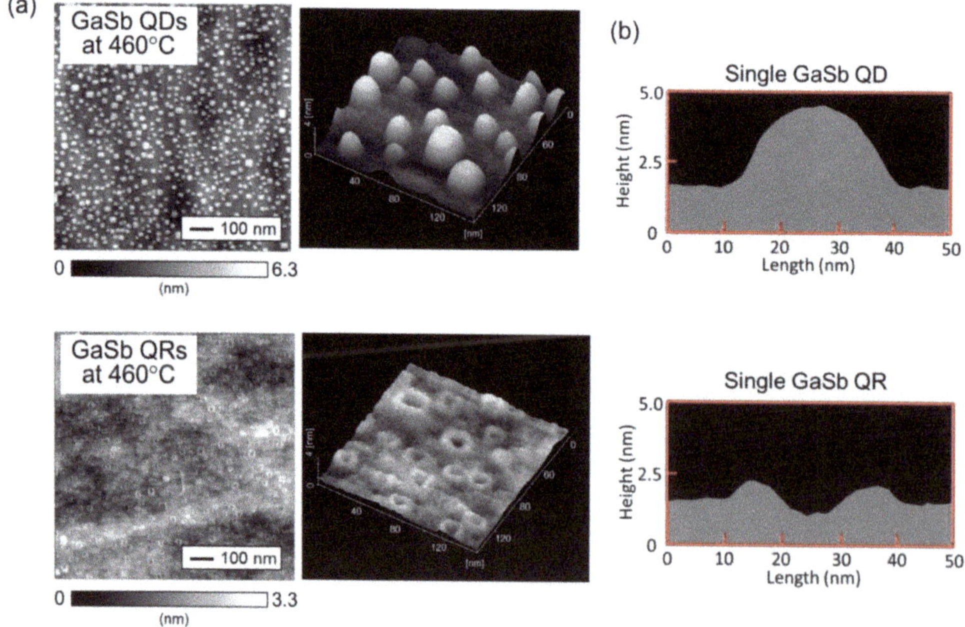

Figure 5. (**a**) Plane and bird's eye views of AFM images of GaSb QDs and QRs grown on an AlGaAs layer at 460 °C. (**b**) Line profiles of a single GaSb QD and QR.

Table 1. Mean diameter, height, and density of GaSb QDs and QRs grown on an AlGaAs layer at 460 °C.

Structure	Diameter (nm)	Height (nm)	Density (cm^{-2})
QDs	27.84	2.56	7.7×10^{10}
QRs	28.44	0.95	7.4×10^{10}

3.2. Characterization of the Device Performances and Carrier Extraction Processes for GaSb/AlGaAs QDSCs and QRSCs

Figure 6 shows the current–voltage curves of the GaSb/AlGaAs QDSCs and QRSCs, which were measured under AM 1.5 G solar spectrum illumination. The measurements were carried out at room temperature. The short-circuit current density (J_{SC}), open-circuit voltage (V_{OC}), and fill factor (FF) of both cells are summarized in Table 2. The trend due to the difference in the quantum nanostructure is the same as that of the GaSb/GaAs system. Owing to the difference in the crystal quality, QRSCs were superior in terms of the device performance compared with QDSCs. Furthermore, both SCs with the AlGaAs host material showed higher V_{OC} compared with the SCs with the GaAs host material [31]. However, in the operation of IBSCs, obtaining a two-step optical transition via IB at higher temperatures is important. Hence, we investigated the temperature dependence of the carrier extraction processes for each SC by measuring EQE and ΔEQE.

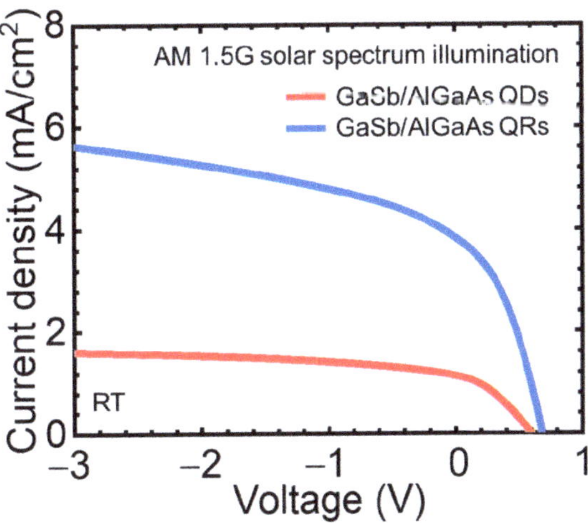

Figure 6. Current–voltage curves for the GaSb/AlGaAs QDSCs and QRSCs measured under AM 1.5 G solar spectrum illumination.

Table 2. Parameters of the light current–voltage curves for the GaSb/AlGaAs QDSCs and QRSCs.

SC	J_{SC} (mA/cm^2)	V_{OC} (V)	FF (-)
GaSb/AlGaAs QD	1.13	0.596	0.351
GaSb/AlGaAs QR	3.83	0.679	0.389

Figure 7 shows (Figure 7a) EQE and (Figure 7b) ΔEQE spectra for GaSb/AlGaAs QDSCs and QRSCs, as obtained at 200 K. For both samples, the EQE responses in the wavelength region from 600 to 740 nm represent the photocurrent production due to the VB–CB optical excitation in the AlGaAs host material. In addition, the EQE signals in the wavelength range longer than 740 nm represent currents produced by the escape of photogenerated carriers in GaSb quantum nanostructures (IB) due to thermal excitation or electric-field assistance. Compared with the QDSC, the QRSC exhibits a higher EQE response associated with the AlGaAs host material. This difference is attributed to the volume of the QR structure being smaller than that of the QD structure, which results in less residual strain during stacking and enables the growth of an AlGaAs layer with fewer crystal defects. In the EQE signals with $\lambda > 740$ nm, the absorption volume of quantum nanostructures and the carrier collection efficiency of each extraction process compete. The ΔEQE signals in the $600 < \lambda < 740$ nm range are associated with the IR-light-induced re-excitation of the holes that relaxed to the IB after being generated in the AlGaAs host material. The ΔEQE signals with $\lambda > 740$ nm are associated with photocurrents produced by the two-step optical excitation via IB. In the wavelength region of absorption by quantum nanostructures, samples with a low EQE and high ΔEQE are better suited for IBSC operation.

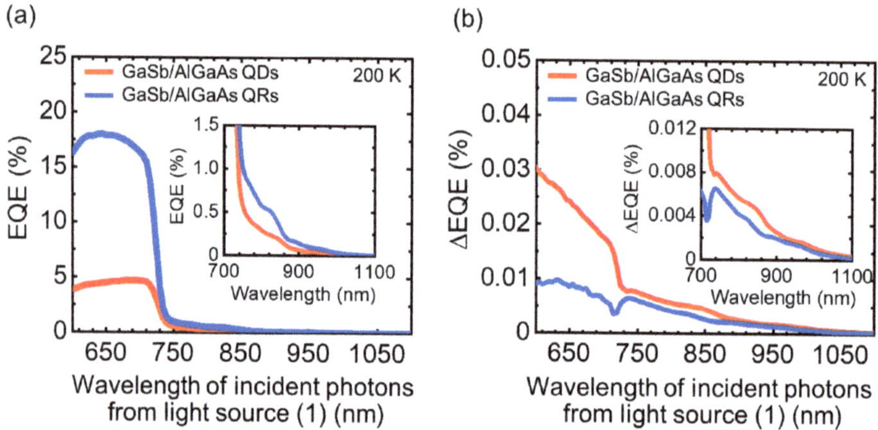

Figure 7. (**a**) EQE and (**b**) ΔEQE spectra for 10-layer stacked GaSb/AlGaAs QDSC and QRSC. The measurement temperature was 200 K. The insets show zoom-in of the photoabsorption region for GaSb quantum nanostructures.

Furthermore, because SCs are, in principle, used at room temperature or higher, current generation due to two-step optical excitation via IB must occur at higher temperatures. Thus, we evaluated the temperature dependence of each photocurrent production process associated with the quantum nanostructures as below. The photocurrent production due to thermal excitation (J_{TE}) and that due to field-assisted escape (J_{FA}) from the IB to the VB were calculated by

$$J_{TE} + J_{FA} = q\int_{\lambda_{CB-VB\_edge}}^{\lambda_{CB-IB\_edge}} \Phi(\lambda)\cdot \mathrm{EQE}(\lambda)\cdot d\lambda, \quad (1)$$

where $\lambda_{CB\text{-}VB\_edge}$ is the photoabsorption edge of the CB–VB determined by the EQE spectrum. $\lambda_{CB\text{-}IB\_edge}$ is the photoabsorption edge of the CB–IB transition and the longest wavelength at which the EQE signal is observed. q is the elementary charge and $\phi(\lambda)$ is the photon flux. The photocurrent produced by optical excitation (J_{OE}) from the IB to the VB was calculated by

$$J_{OE} = q\int_{\lambda_{CB-VB\_edge}}^{\lambda_{CB-IB\_edge}} \Phi(\lambda)\cdot \Delta\mathrm{EQE}(\lambda)\cdot d\lambda, \quad (2)$$

Figure 8a shows the temperature dependence of the sum of J_{TE} and J_{FA} for the GaSb/AlGaAs QDSC and QRSC. In addition, Figure 8a includes data for GaSb/GaAs QDs to compare the effect of the AlGaAs as the host material. The sum of J_{TE} and J_{FA} increased with increasing temperature for all of the samples. In the extremely low temperature region, the J_{FA} is dominant in current production because the thermal excitation is less likely to occur as a result of the low thermal energy. When sufficient thermal energy is generated to enable excitation of the carrier from the IB to the VB, the J_{TE} component is obtained and increases with increasing temperature. For the GaAs/GaSb QDSC, the sum of J_{TE} and J_{FA} increases when the temperature exceeds 100 K, indicating the initiation of thermal excitation. The onset temperature of thermal excitation for the samples with the AlGaAs host material shifts to higher temperatures compared with that for the samples with the GaAs host material. This result indicates that the use of a wide-bandgap material for the barrier layer of quantum nanostructures reduces undesirable carrier escape from the IB to the VB by thermal excitation. Figure 8b shows the temperature dependence of the J_{OE} for the GaSb/GaAs QDSC and the GaSb/AlGaAs QDSC and QRSC. For all of the samples, the J_{OE} decreases with increasing temperature because the component of the thermal extraction amount increases with increasing temperature. Herein, the reduction curve of J_{OE} includes the influence of decreased carrier collection efficiency by an increase in the nonradiative recombination rate because of the increase in temperature. The decrease in J_{OE} with increasing temperature for the GaSb/AlGaAs QDSC was slower than those for the other samples, indicating the feasibility of IBSC operation at higher temperatures.

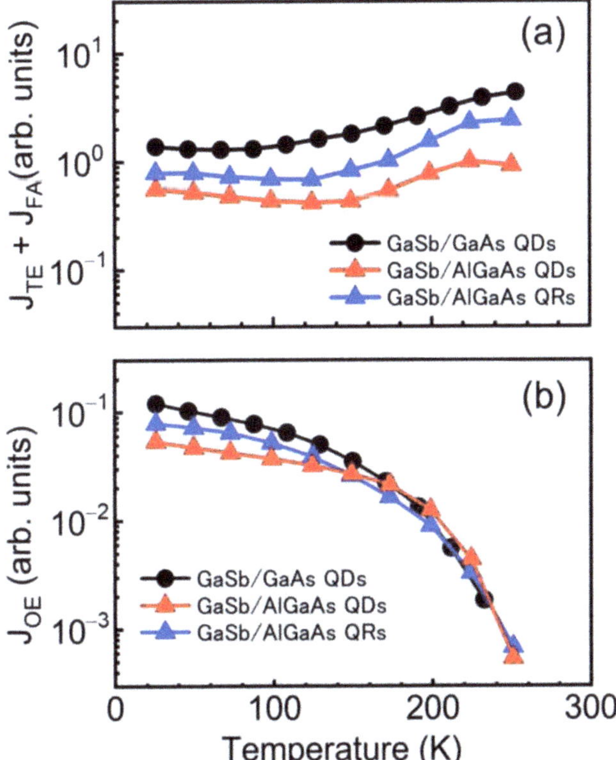

Figure 8. (**a**) Temperature dependence of the J_{TE} and J_{FA} for a 10-layer stacked GaSb/GaAs QDSC, GaSb/AlGaAs QDSC, and GaSb/AlGaAs QRSC. (**b**) Temperature dependence of the J_{OE} for a 10-layer stacked GaSb/GaAs QDSC, GaSb/AlGaAs QDSC, and GaSb/AlGaAs QRSC.

To clarify the suitability for IBSC operation, we evaluated the proportion of the optical excitation component (P_{OE}) in the photocurrent production related to the IB–VB transitions from the following equation:

$$P_{OE} = \frac{J_{OE}}{(J_{OE} + J_{TE} + J_{FA})} \quad (3)$$

A higher value of P_{OE} means a more suitable structure for IBSC. Figure 9 shows the temperature dependence of P_{OE} for the GaSb/GaAs QDSC and the GaSb/AlGaAs QDSC and QRSC.At lower temperatures, the P_{OE} is higher for samples with the AlGaAs host material than for the GaSb/GaAs QDSC. The P_{OE} curve of the GaSb/GaAs QDSC decreases substantially with increasing temperature; when the temperature is increased to 150 K, the value decreases to one-fourth of that at 26 K. By contrast, the GaSb/AlGaAs QDSC maintained a P_{OE} value of 6% even at 150 K. A higher energy was required for carrier escape by thermal excitation in the IB–VB transition for samples with the AlGaAs host material. Thus, carrier extraction by optical excitation occurred even at high temperatures. Moreover, the temperature characteristic of the P_{OE} for the GaSb/AlGaAs QDSC was superior to that of the P_{OE} for the GaSb/AlGaAs QRSC. This difference is attributed to the confinement strength of the quantum nanostructure. As shown in Table 1, the height of the QR structure is only 0.95 nm, whereas the height of the QD structure is 2.56 nm. Thus, compared with the IB in the QD structure, that in the QR structure is formed at a higher energy position within the potential. The results demonstrate that the QDSC prepared with wide-bandgap host-layer materials are suitable for IBSC operation. However, GaSb QDs tend to form defects in the host material because of the growth of multilayer structures; therefore, further research into growth conditions and structures (e.g., introduction of an interlayer between GaSb and AlGaAs [20–22]) that result in high-quality QDSCs is needed.

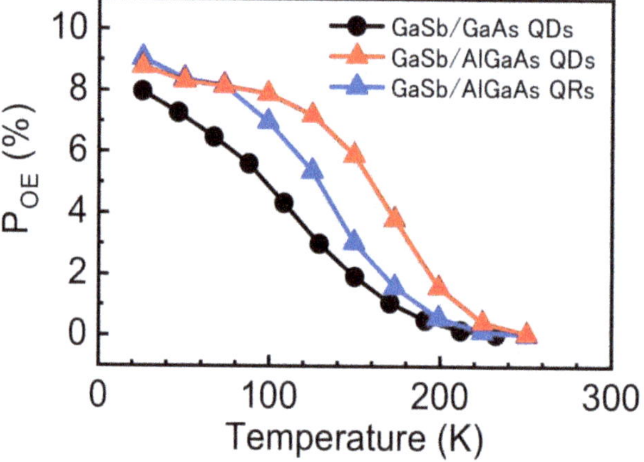

Figure 9. Temperature dependence of the P_{OE} for a 10-layer stacked GaSb/GaAs QDSC, GaSb/AlGaAs QDSC, and GaSb/AlGaAs QRSC.

4. Conclusions

We studied 10-layer stacked GaSb/AlGaAs QDSCs and QRSCs for use in IBSCs. GaSb QDs and QRs were grown on an AlGaAs layer. We demonstrated that GaSb QDs and QRs can be formed on AlGaAs layers by Stranski–Krastanov growth and As-soaking techniques. Further, we evaluated the suitability of the GaAs/GaAs QDSC and the GaSb/AlGaAs QDSC and QRSC for the IBSC using the temperature characteristics of the P_{OE} value. By changing the host material from GaAs to AlGaAs, a high P_{OE} value was obtained. Among the samples in the present study, the GaSb/AlGaAs QDSC showed the best temperature characteristics with respect to the P_{OE} value. This result is attributed to the formation of a strong quantum confinement structure in the GaSb/AlGaAs QDSC, which suppresses the unintended escape of carriers in the IB as a result of thermal excitation.

Author Contributions: Conceptualization, Y.S. and Y.O.; methodology, Y.S. and R.T.; analysis, Y.S. and R.T.; investigation, Y.S. and R.T.; writing—original draft preparation, Y.S.; writing—review and editing, R.T. and Y.O.; and supervision, Y.O. All authors have read and agreed to the published version of the manuscript.

Funding: This research was supported by the New Energy and Industrial Technology Development Organization (NEDO) and Grant-in-Aid for Early-Career Scientists (JP20K15183) from the Japan Society for the Promotion of Science (JSPS).

Conflicts of Interest: The authors declare no conflict of interest.

References

1. Shockley, W.; Queisser, H.J. Detailed balance limit of efficiency of p-n junction solar cells. *J. Appl. Phys.* **1961**, *32*, 510–519. [CrossRef]
2. Luque, A.; Martí, A. Increasing the Efficiency of Ideal Solar Cells by Photon Induced Transitions at Intermediate Levels. *Phys. Rev. Lett.* **1997**, *78*, 5014–5017. [CrossRef]
3. Luque, A.; Martí, A.; Stanley, C. Understanding intermediate-band solar cells. *Nat. Photonics* **2012**, *6*, 146–152. [CrossRef]
4. Okada, Y.; Ekins-Daukes, N.J.; Kita, T.; Tamaki, R.; Yoshida, M.; Pusch, A.; Hess, O.; Phillips, C.C.; Farrell, D.J.; Yoshida, K.; et al. Intermediate band solar cells: Recent progress and future directions. *Appl. Phys. Rev.* **2015**, *2*, 021302. [CrossRef]
5. Bae, D.; Kanellos, G.; Wedege, K.; Draževič, E.; Bentien, A.; Smith, A.W. Tailored energy level alignment at MoO_X/GaP interface for solar-driven redox flow battery application. *J. Chem. Phys.* **2020**, *152*, 124710. [CrossRef] [PubMed]
6. Hubbard, S.M.; Cress, C.D.; Bailey, C.G.; Raffaelle, R.P.; Bailey, S.G.; Wilt, D.M. Effect of strain compensation on quantum dot enhanced GaAs solar cells. *Appl. Phys. Lett.* **2008**, *92*, 123512. [CrossRef]
7. Okada, Y.; Oshima, R.; Takata, A. Characteristics of InAs/GaNAs strain-compensated quantum dot solar cell. *J. Appl. Phys.* **2009**, *106*, 024306. [CrossRef]
8. Popescu, V.; Bester, G.; Hanna, M.C.; Norman, A.G.; Zunger, A. Theoretical and experimental examination of the intermediate-band concept for strain-balanced (In,Ga)As/Ga(As,P) quantum dot solar cells. *Phys. Rev. B* **2008**, *78*, 205321. [CrossRef]
9. Sugaya, T.; Numakami, O.; Oshima, R.; Furue, S.; Komaki, H.; Amano, T.; Matsubara, K.; Okano, Y.; Niki, S. Ultra-high stacks of InGaAs/GaAs quantum dots for high efficiency solar cells. *Energy Environ. Sci.* **2012**, *5*, 6233–6237. [CrossRef]
10. Tanabe, K.; Guimard, D.; Bordel, D.; Arakawa, Y. High-efficiency InAs/GaAs quantum dot solar cells by metalorganic chemical vapor deposition. *Appl. Phys. Lett.* **2012**, *100*, 193905. [CrossRef]
11. Tamaki, R.; Shoji, Y.; Okada, Y.; Miyano, K. Spectrally Resolved Interband and Intraband Transitions by Two-Step Photon Absorption in InGaAs/GaAs Quantum Dot Solar Cells. *IEEE J. Photovolt.* **2015**, *5*, 229–233. [CrossRef]
12. Kada, T.; Asahi, S.; Kaizu, T.; Harada, Y.; Kita, T.; Tamaki, R.; Okada, Y.; Miyano, K. Two-step photon absorption in InAs/GaAs quantum-dot superlattice solar cells. *Phys. Rev. B* **2015**, *91*, 201303. [CrossRef]
13. Hirao, K.; Asahi, S.; Kaizu, T.; Kita, T. Two-step photocurrent generation enhanced by the fundamental-state miniband formation in intermediate-band solar cells using a highly homogeneous InAs/GaAs quantum-dot superlattice. *Appl. Phys. Express* **2018**, *11*, 012301. [CrossRef]
14. Shoji, Y.; Tamaki, R.; Datas, A.; Martí, A.; Luque, A.; Okada, Y. Effect of field damping layer on two step absorption of quantum dots solar cells. In Proceedings of the Technical Digests of 6th World Conference on Photovoltaic Energy Conversion, Kyoto, Japan, 23–27 November 2014.
15. Datas, A.; López, E.; Ramiro, I.; Antolín, E.; Martí, A.; Luque, A.; Tamaki, R.; Shoji, Y.; Sogabe, T.; Okada, Y. Intermediate Band Solar Cell with Extreme Broadband Spectrum Quantum Efficiency. *Phys. Rev. Lett.* **2015**, *114*, 157701. [CrossRef] [PubMed]
16. Linares, P.G.; López, E.; Ramiro, I.; Datas, A.; Antolín, E.; Shoji, Y.; Sogabe, T.; Okada, Y.; Martí, A.; Luque, A. Voltage limitation analysis in strain-balanced InAs/GaAsN quantum dot solar cells applied to the intermediate band concept. *Sol. Energy Mater. Sol. Cells* **2015**, *132*, 178–182. [CrossRef]
17. Ramiro, I.; Antolín, E.; Steer, M.; Linares, P.; Hernandez, E.; Artacho, I.; Lopez, E.; Ben, T.; Ripalda, J.; Molina, S. InAs/AlGaAs quantum dot intermediate band solar cells with enlarged sub-bandgaps. In Proceedings of the 38th IEEE Photovoltaic Specialists Conference, Austin, TX, USA, 3–8 June 2012.
18. Tamaki, R.; Shoji, Y.; Sugaya, T.; Okada, Y. Design optimization for two-step photon absorption in quantum dot solar cells by using infrared photocurrent spectroscopy. In Proceedings of the SPIE 9743, Physics, Simulation, and Photonic Engineering of Photovoltaic Devices V, San Francisco, CA, USA, 14 March 2016; p. 974318.
19. Hosokawa, H.; Tamaki, R.; Sawada, T.; Okonogi, A.; Sato, H.; Ogomi, Y.; Hayase, S.; Okada, Y.; Yano, T. Solution-processed intermediate-band solar cells with lead sulfide quantum dots and lead halide perovskites. *Nat. Commun.* **2019**, *10*, 43. [CrossRef]
20. Sugaya, T.; Oshima, R.; Matsubara KNiki, N. InGaAs quantum dot superlattice with vertically coupled states in InGaP matrix. *J. Appl. Phys.* **2013**, *114*, 014303. [CrossRef]
21. Ramiro, I.; Villa, J.; Lam, P.; Hatch, S.; Wu, J.; López, E.; Antolín, E.; Liu, H.; Martí, A. Wide-Bandgap InAs/InGaP Quantum-Dot Intermediate Band Solar Cells. *IEEE J. Photovolt.* **2015**, *5*, 840–845. [CrossRef]
22. Lam, P.; Wu, J.; Tang, M.; Kim, D.; Hatch, S.; Ramiro, I.; Dorogan, V.G.; Benamara, M.; Mazur, Y.I.; Salamo, G.J.; et al. InAs/InGaP quantum dot solar cells with an AlGaAs interlayer. *Sol. Energy Mater. Sol. Cells* **2016**, *144*, 96–101. [CrossRef]
23. Laghumavarapu, R.B.; Moscho, A.; Khoshakhlagh, A.; El-Emawy, M.L.; Lester, F.; Huffaker, D.L. GaSb/GaAs type II quantum dot solar cells for enhanced infrared spectral response. *Appl. Phys. Lett.* **2007**, *90*, 173125. [CrossRef]
24. Tayagaki, T.; Sugaya, T. Type-II InP quantum dots in wide-bandgap InGaP host for intermediate-band solar cells. *Appl. Phys. Lett.* **2016**, *108*, 153901. [CrossRef]

25. Kim, D.; Hatch, S.; Wu, J.; Sablon, K.A.; Lam, P.; Jurczak, P.; Tang, M.; Gillin, W.P.; Liu, H. Type-II InAs/GaAsSb Quantum Dot Solar Cells With GaAs Interlayer. *IEEE J. Photovolt.* **2018**, *8*, 741–745.
26. Suzuki, R.; Terada, K.; Sakamoto, K.; Sogabe, T.; Yamaguchi, K. Low sunlight concentration properties of InAs ultrahigh-density quantum-dot solar cells. *Jpn. J. Appl. Phys.* **2019**, *58*, 071004. [CrossRef]
27. Ramiro, I.; Villa, J.; Hwang, J.; Martin, A.J.; Millunchick, J.; Phillips, J.; Martí, A. Demonstration of a GaSb/GaAs Quantum Dot Intermediate Band Solar Cell Operating at Maximum Power Point. *Phys. Rev. Lett.* **2020**, *125*, 247703. [CrossRef] [PubMed]
28. Kechiantz, A.; Afanasev, A.; Lazzari, J. Impact of spatial separation of type-II GaSb quantum dots from the depletion region on the conversion efficiency limit of GaAs solar cells. *Prog. Photovolt. Res. Appl.* **2015**, *23*, 1003–1016. [CrossRef]
29. Timm, R.; Lenz, A.; Eisele, H.; Ivanova, L.; Dähne, M.; Balakrishnan, G.; Huffaker, D.L.; Farrer, I.; Ritchie, D.A. Quantum ring formation and antimony segregation in GaSb/GaAs nanostructures. *J. Vac. Sci. Technol. B* **2008**, *26*, 1492. [CrossRef]
30. Lin, S.; Lin, W.; Tseng, C.; Chen, S. GaSb/GaAs quantum dots with type-II band alignments prepared by molecular beam epitaxy for device applications. In Proceedings of the SPIE OPTO 7947, 79470M–2, San Francisco, CA, USA, 1 March 2011.
31. Shoji, Y.; Tamaki, R.; Okada, Y. Multi-stacked GaSb/GaAs type-II quantum nanostructures for application to intermediate band solar cells. *AIP Adv. Lett.* **2017**, *7*, 065305. [CrossRef]
32. Mo, Y.W.; Savage, D.E.; Swartzentruber, B.S.; Lagally, M.G. Kinetic pathway in Stranski-Krastanov growth of Ge on Si (001). *Phys. Rev. Lett.* **1990**, *65*, 1020. [CrossRef]
33. Baskaran, A.; Smerek, P. Mechanisms of Stranski-Krastanov growth. *J. Appl. Phys.* **2012**, *111*, 044321.
34. Wang, M.W.; Collins, D.A.; McGill, T.C.; Grant, R.W.; Feenstra, R.M. Study of interface asymmetry in InAs–GaSb heterojunctions. *J. Vac. Sci. Technol. B* **1995**, *13*, 1689–1693. [CrossRef]

Article

Nanostructure of Porous Si and Anodic SiO_2 Surface Passivation for Improved Efficiency Porous Si Solar Cells

Panus Sundarapura [1], Xiao-Mei Zhang [1,2,*], Ryoji Yogai [3], Kazuki Murakami [3], Alain Fave [4] and Manabu Ihara [1,3,*]

1 Department of Chemical Science and Engineering, Tokyo Institute of Technology, 2-12-1 Ookayama, Meguro, Tokyo 152-8552, Japan; sundarapura.p.aa@m.titech.ac.jp
2 Department of Mechanical Engineering, Tokyo Institute of Technology, 2-12-1 Ookayama, Meguro, Tokyo 152-8552, Japan
3 Department of Chemistry, Tokyo Institute of Technology, 2-12-1 Ookayama, Meguro, Tokyo 152-8552, Japan; energy.power.ry.516@gmail.com (R.Y.); kazuki.techtech@gmail.com (K.M.)
4 Univ Lyon, INSA Lyon, ECL, CNRS, UCBL, CPE Lyon, INL, UMR5270, 69621 Villeurbanne, France; alain.fave@insa-lyon.fr
* Correspondence: xiaomeizhang2007@gmail.com (X.-M.Z.); mihara@chemeng.titech.ac.jp (M.I.)

Abstract: The photovoltaic effect in the anodic formation of silicon dioxide (SiO_2) on porous silicon (PS) surfaces was investigated toward developing a potential passivation technique to achieve high efficiency nanostructured Si solar cells. The PS layers were prepared by electrochemical anodization in hydrofluoric acid (HF) containing electrolyte. An anodic SiO_2 layer was formed on the PS surface via a bottom-up anodization mechanism in HCl/H_2O solution at room temperature. The thickness of the oxide layer for surface passivation was precisely controlled by adjusting the anodizing current density and the passivation time, for optimal oxidation on the PS layer while maintaining its original nanostructure. HRTEM characterization of the microstructure of the PS layer confirms an atomic lattice matching at the PS/Si interface. The dependence of photovoltaic performance, series resistance, and shunt resistance on passivation time was examined. Due to sufficient passivation on the PS surface, a sample with anodization duration of 30 s achieved the best conversion efficiency of 10.7%. The external quantum efficiency (EQE) and internal quantum efficiency (IQE) indicate a significant decrease in reflectivity due to the PS anti-reflection property and indicate superior performance due to SiO_2 surface passivation. In conclusion, the surface of PS solar cells could be successfully passivated by electrochemical anodization.

Keywords: solar cell; porous silicon; surface passivation; anodic oxidation

Citation: Sundarapura, P.; Zhang, X.-M.; Yogai, R.; Murakami, K.; Fave, A.; Ihara, M. Nanostructure of Porous Si and Anodic SiO_2 Surface Passivation for Improved Efficiency Porous Si Solar Cells. *Nanomaterials* **2021**, *11*, 459. https://doi.org/10.3390/nano11020459

Academic Editor: Zafar Iqbal

Received: 30 December 2020
Accepted: 1 February 2021
Published: 11 February 2021

Publisher's Note: MDPI stays neutral with regard to jurisdictional claims in published maps and institutional affiliations.

1. Introduction

Solar cells, as a renewable energy source, will significantly contribute to clean energy, which is an energy goal that has become one of the greatest challenges to fuel our economies. For solar cells, silicon (Si) is a promising material that combines suitable optoelectronic properties with earth-abundance and technological availability [1]. Si solar cells are dominating the photovoltaic (PV) market, in which state-of-the-art *c*-Si solar cells can achieve about 26.7% efficiency [2], approaching the Shockley–Queisser limit of 30% [3]. In order to surpass the theoretical efficiency limit for single junction solar cells, a tandem structure is a potential candidate. By selecting the material based on the band gap energy by considering the atomic lattice matching, nanostructured silicon material could be a promising candidate for the top layer of the tandem structure. For decades, there has been extensive research into potential candidates for the top layer, and the most prominent Si-based materials are Si nanowire [4] and nanopore arrays [5,6]. These structures are composed of an array of vertically aligned Si nanostructures, which can not only reduce the volume of semiconductor absorber but can provide nearly ideal photon absorption across the solar spectrum for high efficiency. When the pore size of the surface features equals or exceeds the wavelength of the incident light, nanostructures provide beneficial light-trapping that increases the effective light-path length in the Si [7,8]. When its porosity is increased, porous Si (PS) exhibits decreased reflectance (3~30%) and increased band gap (1.1~1.9 eV) [9]. Such unique features make PS a promising material for use in solar cell technology and an interesting candidate for the top absorber in Si tandem cell applications [10].

As high aspect-ratio nanostructures, like porous silicon, are prone to an uncontrollable oxidation of native oxide when exposed to the atmosphere, many studies have been conducted on various passivation techniques to improve their porous surface stability, e.g., wet thermal oxidation [11], rapid thermal treatment in NH_3 [12], and room temperature oxidation with ozone [13]. However, as these techniques only applied to the PS layer without *p-n* junction, their applications are only limited to PS as the photoluminescence (PL) emitting layer for other applications, like biosensors and microelectronics. In terms of applications in PV cells, compared with conventional cells in the present time, solar cells based on nanostructured Si [14–16] exhibit lower efficiencies without surface passivation. The main challenges that hamper the application of PS in solar cells are related to the increased photocarriers at the dramatically increased surface area of nano-porous Si causing reductions in both the device's short-circuit current and open-circuit voltage, and consequently causing a decrease in the efficiency (η). This indicates that the passivation of PS surface is important to improve the performance of the nanostructured solar cell.

In terms of passivation of the surface of Si, Al_2O_3 films deposited on Si surfaces via atomic layer deposition (ALD) possess negative fixed charges at the Al_2O_3/Si interface, and thus provide superior passivation quality for p^+ emitters of *n*-type-based solar cells [17,18]. For phosphorus diffused n^+ emitters of *p*-type solar cells, SiO_2 films can provide good passivation due to their excellent chemical passivation for the Si-SiO_2 interface [19]. The recent study by Grant et al. demonstrates that the thin SiO_2 film (~70 nm) formed by the anodic oxidation method at room temperature can exhibit an excellent surface passivation quality and durability on the *c*-Si substrate with a high η of >23% [20]. However, the anodic oxidation was not applied to the nanostructured solar cell. In the case of nanostructured solar cells, additional technological issues, such as the controllability of SiO_2 thickness, the effectiveness of passivation by the nanometer scale thin SiO_2 layer, and the preservation of the original structure, must be developed. Oh et al. fabricated a *p*-type-based black Si solar cell in which a thermal SiO_2 passivation layer provided a very low effective surface recombination velocity of 53 cm s^{-1} on the surface and yielded an η of 18.2% [21]. However, the process to thermally grow SiO_2, despite having a small effect on the structure when applied to *c*-Si, has a limitation when applied to PS nanostructures that have an aspect ratio as high as 400:1 [22]. The high operating temperature of thermal SiO_2 deposition at 400–1100 °C [11,12] will damage the nano-porous layer [23]. Therefore, the method to deposit an oxide passivation layer on nanostructured Si must be designed in such a way that it can precisely control the thickness of the oxide to prevent a complete oxidation, while the processing temperature must be low enough to preserve the original nanostructure.

In this work, we used a simple, low-cost method of electrochemical etching to prepare PS layers with a *p-n* junction with low reflectivity on *n*-Si emitters of Si solar cells. The PS was composed of crystalline nano-pillars and possessed low reflectivity in the ultraviolet (UV) and visible regimes (300–1100 nm). Electrochemical anodization was applied to passivate the surface of the PS layer because this passivation technique allows the precise control of the oxide layer thickness and can be applied at room temperature in a short processing time, which also reduces the risk of adversely affecting the original nanostructure of PS compared to conventional high temperature passivation processes. This low-cost approach combines both the good surface uniformity of SiO_2 and the effective oxidation of the dangling bond interface states to provide sufficient passivation for PS solar cells. These PS solar cells could efficiently reduce the recombination at the front surface via a bottom-up anodization mechanism used to form the SiO_2 passivation layer on the surface of the pores. The effect of the anodization duration on the main features of the PV devices was demonstrated systematically by evaluating the PV performance of these anodic SiO_2 passivated PS solar cells. In these PS solar cells, an optimum anodization duration of 30 s produced a maximum η of 10.7%, associated with passivation time (t) effects on the open-circuit voltage (V_{oc}), short-circuit current density (J_{sc}), fill factor (FF), shunt resistance (R_{sh}), and series resistance (R_s). The functionality of the PS layer for improving the PV performance of solar cells was investigated here by varying the SiO_2 layer thickness.

With the existing technologies to fabricate the PS layer in more scalable production [24], e.g., in microelectronics [25] and high-volume industrial microelectromechanical systems (MEMS) production [26], together with the reduction in the manufacturing cost by the double layer transfer process (DLPS) method that allowed the utilization of the silicon wafers in a more efficient way [6,27,28], the fabrication of PS as well as the SiO_2 passivation could implement these technologies, with some modifications, in the actual PV production scale. Therefore, as the critical step to develop a more feasible passivation technique for PS solar cells, the electrochemical passivation of PS as the top emitter of Si solar cell at room temperature highlights a promising possibility that this low-cost process for PS structure can improve the light-trapping in thin-film Si solar cells and possibly be used as a wide band gap, top-cell material in Si tandem cell applications in the next generation solar cells.

2. Materials and Methods

The substrates of the PS solar cells fabricated here were p-type c-Si wafers with (100) orientation, a resistivity of 1–2 Ω cm, a thickness of 200 μm and an area of about 177 cm^2. The Si wafers were ultrasonically cleaned and then dipped in dilute HF 1% (Wako Pure Chemical Industries, Ltd., Osaka, Japan) solution to remove the native oxide. A phosphorus (P) doped n layer was deposited on the p-Si wafer surface via spin-on doping (SOD Standard Solution P8545PV, Filmtronics, Butler, PA, USA) with post-annealing at 900 °C (MILA-5000 series, Advance Riko, Inc., Yokohama, Japan) for 17 min under a nitrogen atmosphere. To characterize the doping profile and the p-n interface of this fabrication condition, the electrochemical capacitance voltage (ECV) (WEP Wafer Profiler CVP21, Furtwangen, Germany) was measured for four samples (Samples A,B,C,D; Table 1) of phosphorus (P) doped bulk p-Si wafers after removal of the thermal oxide from the wafer surface. Repeated measurements were performed for three samples that had a 5 mm^2 etching area (Samples A-C) and for three different areas (01–03) on the same sample (D). Note that this etching process mentioned in the "etching area" is an electrolytic etching of the doped c-Si only for profiling the carrier concentration in each depth from the surface, and differs from the process of fabricating the PS layer in the later section.

To fabricate the PS layer, an anodization process was performed using an in-house constructed wet-etching cell, as shown in Figure 1a, with horizontal orientation of the Si wafers [6,29]. The lower electrical contact to the wafers was provided via a copper base. A cylindrical cell containing the electrolyte of 46% HF: 99.5% ethanol = 1:1 solution was placed on top of the Si wafer. The surface area size of the PS layer formed on the Si wafer surface was controlled using an O-ring with an inner diameter of 20 mm. The cathode was a platinum net (15 mm in diameter) immersed in the electrolyte. Etching was performed under dark conditions at room temperature. Furthermore, for the samples with surface passivation, the formation of anodic SiO_2 was carried out by in-situ surface passivation of the PS layer using an anodization process in a HCl/H_2O solution at a constant current of 0.36 mA/cm^2 for different passivation time t from 6 s to 135 s.

Table 1. Overview of preparation conditions and experiments for all samples.

Samples	Current Density (mA/cm^2)	Etching Duration (s)	Passivation Duration (s)	PS Layer Thickness (nm)	Volume Ratio of SiO_2 to PS Layer (%)	Experiment	Section
Samples A,B,C	0	0	0	0	0	ECV 5 mm^2	Results and Discussion part 3.1
Sample D	0	0	0	0	0	ECV 0.7 mm^2	
PS-1 layer	75	6	0	146	0	TEM HRTEM	
PS-2 layer	75	6	0	176	0		
PS-3 layer	① 75 ② 5	① 10 ② 300	0	545	0		
c-Si cell	0	0	0	0	0	Reflectivity *J–V* curve EQE	Results and Discussion part 3.2
As-fabricated PS-1 cell	75	6	0	150	0		
Passivated PS-1 cell	75	6	-	-	-		
As-fabricated PS-2 cell	75	6	0	191	0	FESEM Reflectivity *J–V* curve EQE IQE	Results and Discussion part 3.3
Passivated PS-2-a cell	75	6	6	239	0.6		
Passivated PS-2-b cell	75	6	10	224	0.9		
Passivated PS-2-c cell	75	6	15	205	1.4		
Passivated PS-2-d cell	75	6	20	255	1.9		
Passivated PS-2-e cell	75	6	30	235	2.8		
Passivated PS-2-f cell	75	6	60	216	5.6		
Passivated PS-2-g cell	75	6	90	213	8.5		
Passivated PS-2-h cell	75	6	135	210	12.7		

[-: data not available].

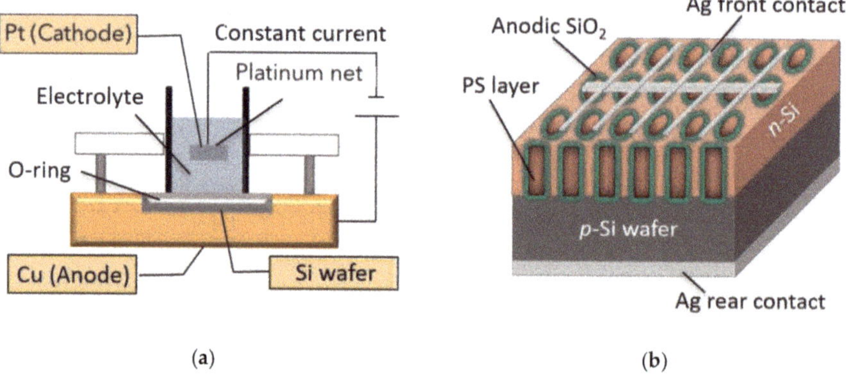

Figure 1. Schematic of (**a**) electrochemical etching set-up; (**b**) solar cells based on porous silicon (PS).

With this fabrication condition, PS layers of different thicknesses and an average porosity of 38% were prepared on *c*-Si substrates. A porosity of 38% was calculated based on the effective medium approximation (EMA) with Bruggeman's Model [30] and is optimal to achieve a low lattice-mismatch between the PS layers and *c*-Si substrate. For the surface passivation, a thin SiO_2 film was formed to cover the surfaces of the pores on the PS layer by the anodization method in H_2O/HCl at ambient temperature. The anodization reaction is as follows:

$$Si + 2H_2O \leftrightarrow SiO_2 + 2H_2 \quad (1)$$

Based on this porosity and the kinetic of the anodization reaction (1) [31], the volume ratio of SiO_2 thin film per volume of PS layer in each t can be calculated, so that it can be confirmed that the PS layer is not completely oxidized, i.e., the volume ratio is equal to 100%. Table 1 summarizes the preparation conditions and layer thickness of all the samples fabricated for this study.

Transmission electron microscopy (TEM, 200 KV JEM-2010F, JEOL, Tokyo, Japan) was used to investigate the microstructures of the PS-1 layer, PS-2 layer and PS-3 layer. TEM specimens were prepared using a focus ion beam (FIB) (FB-2100) procedure. A protective 200-nm-thick coating of carbon was deposited on the surface of the PS layer. A foil from the PS layer was obtained using Ga+ ions beam to excavate the material from both sides to a depth of 5 μm. Finally, the foil was extracted from the PS layer and deposited onto the TEM membrane. The morphology of the PS cells was characterized using a Field Emission Scanning Electron Microscope (FESEM, JSM-6301FZ, JEOL, Tokyo, Japan). FESEM samples were taken from cut samples of the PS-2 cell and passivated PS-2 cells that had been Pt coated (Platinum Auto Fine Coater JFC-1600, JEOL, Tokyo, Japan) for better conductivity in this FESEM imaging.

Finally, to evaluate the PV performance of the fabricated cells, for an active area of 1 cm^2 on the as-fabricated PS-1 cell, passivated PS-1 cell, PS-2 cell, and passivated PS-2-(a-h) cells, the contact electrodes (50-nm-thick Ag films) were deposited with in-house resistive thermal evaporator onto both the front and rear surfaces, as shown in Figure 1b. A fishbone-shaped electrode structure with a 0.39 cm^2 area was used for the front contact. Note that the thin Ag electrode in this experiment has a low transmittance on the wavelength of 300–1200 nm, which is the range that *c*-Si solar cell and PS solar cell absorb the light. Thus, the PV parameter can be calculated based on only the active area. In addition, because it is needed to evaluate only the true effectiveness of the anodic SiO_2 passivation layer, this fabrication method was designed to be as simple as possible, e.g., without back surface field (BSF). The structure of the PS solar cell would be as follows (from top to bottom):

Ag front electrode/anodic SiO_2/N+ emitter PS/ P-type *c*-Si/Ag back electrode.

Then, the current density–voltage (J–V) curve and other PV parameters (J_{sc}, V_{oc}, FF, and η) were obtained using a solar simulator (CEP-25MLH Spectrometer, Bunkoukeiki, Tokyo, Japan) under AM1.5 illumination (Xenon lamp model XCS-150A, JASCO, Tokyo, Japan). The EQE data were obtained using a spectral response measurement system (CEP-2000ML, Bunkoukeiki, Tokyo, Japan), while the IQE data were calculated based on the reflectance data obtained using UV-Vis spectrophotometers (V-570, JASCO, Tokyo, Japan). Note that the J_{sc}, η, EQE and IQE shown in the Results and Discussion section were calculated based on the active area of the solar cell surface. As reference, *c*-Si solar cells were fabricated under the same conditions as those for PS solar cells.

3. Results and Discussion

3.1. Characterization of Nanostructured PS and Depth of P-N Interface

To investigate the depth of the *p*-*n* interface, ECV profiling was performed on phosphorus (P) doped bulk *p*-Si wafers after removal of the thermal oxide from the wafer surface (Samples A,B,C,D; Table 1), to obtain the activated P concentration profiles at 900 °C for 17 min.

In both the arithmetic scale and logarithmic scale (Figure 2), the profiles of these samples show a steep drop-off in the carrier concentration around the wafer depth of 200 nm, which indicates a high P-doping concentration of the *n*-layer on the top of the *p*-Si wafer. Despite the difference of the carrier concentration between samples from the top surface to around 150 nm depth (Figure 2a), it does not significantly affect the position of the *p*-*n* junction, as the difference in this highly doped sample is almost similar, as can be seen in the logarithmic scale (Figure 2b), and the carrier concentration drastically drops down in the same manner in every sample. Repeated measurements from different samples (A–D) and from areas 01–03 from sample D confirm that a uniform *n*-layer had formed on the top of *p*-Si wafer by thermal diffusion of the P-dopant. Figure 2c shows a cross-sectional

TEM image of the 545-nm-thick PS-3 layer on a *c*-Si substrate, revealing bright and dark contrast layers due to their different porosity. The porosity in the PS-3 layer formed on the top bright layer (200-nm-thick) differed from that in the PS-3 layer on the bottom dark layer (270-nm-thick), indicating a different P-doping density of the Si *n*-type region. The porosity of PS can, therefore, be controlled by adjusting the P-doping level [32]. Note that the depth of the top bright contrast layer coincides with the high P-doping concentration of the *n*- layer indicated in the ECV profile.

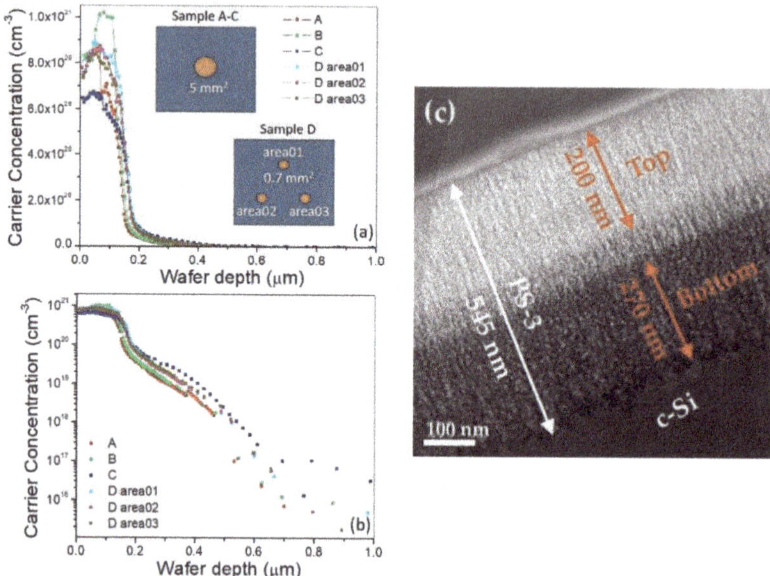

Figure 2. Carrier density dependence on wafer depth by ECV profiling on phosphorus (P) doped bulk Si Samples A–C ($5\ mm^2$ etching size) and Sample D areas 01–03 ($0.7\ mm^2$ etching size): (**a**) arithmetic scale; (**b**) logarithmic scale; (**c**) cross-sectional TEM image of 545-nm-thick PS-3 layer.

Figure 3a shows an overview of the PS-1 layer with a uniform thickness of 146 nm on a *c*-Si substrate. The pores appear as an amorphous material due to amorphization of the thin Si walls around the pores. The thinness of the Si walls was caused by FIB milling used to prepare the sample for TEM imaging. The projection of the Si skeleton is similar in appearance to a crystalline material. The PS-1 layer exhibits straight pores grown perpendicular to the surface. Uniform contrast in the PS-1 layer indicates uniform porosity, which in turn indicates a uniform doping concentration of the *n*-layer. A high resolution TEM (HRTEM) image of the PS-1/*c*-Si interface is shown in Figure 3b. The crystalline skeleton of the *n*-PS-1 layer has the same orientation as the *c*-Si substrate, indicating an atomic lattice matching (marked by red dotted lines in Figure 3b) at the PS-1/Si interface for homojunction Si tandem-cell applications. Figure 3c shows the 176-nm-thick PS-2 layer. This observed layer thickness (176 nm) is less than the as-fabricated 200-nm-thick PS-2 layer due to a slight top-surface etching effect from the FIB milling pretreatment process for TEM imaging. The corresponding HRTEM image (Figure 3d) shows a smooth PS-2/*c*-Si interface. The speckled contrast of dark and bright regions represents Si crystallites and pores, respectively. In the top PS-2 layer, the pore size is about 10 nm with a mean distance between neighboring pores of about 15 nm, while the bottom PS-2 layer shows a smaller pore size of 5 nm with a neighboring distance of 8 nm.

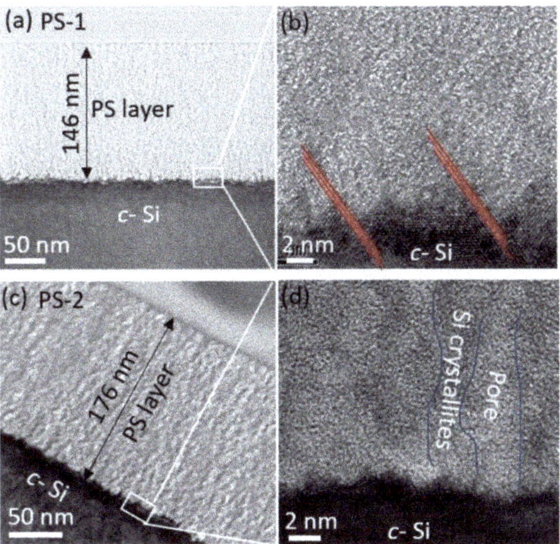

Figure 3. (**a**) TEM image of 146-nm-thick PS-1 layer; (**b**) HRTEM of PS-1/c-Si interface at the bottom of PS-1 layer; (**c**) TEM image of 176-nm-thick PS-2 layer; (**d**) HRTEM of PS-2/c-Si interface at the bottom of PS-2 layer.

3.2. Electrochemical Passivation Effect in Solar Cells with Thin PS Layer of 146 nm

The reflectivity spectrum for the PS-1 sample with the thin 146-nm PS layer (Figure 4a) shows a significantly lower reflectivity compared to that of the bulk Si sample in the wavelength range of 300–1100 nm measured using a UV-Vis spectrophotometer. For this PS-1 sample, the lowest reflectivity of approximately 9% was achieved for the PS-1 layer at a wavelength of 700–1100 nm, compared to 18% at 450–700 nm. This low reflectivity demonstrates the potential anti-reflection property of the PS-1 porous layer. Figure 4b shows the *J*–*V* characteristics for this PS-1 cell, passivated PS-1 cell, and the reference *c*-Si cell, measured under AM1.5 illumination. Note that there was no other anti-reflection coating for any of the solar cells in this study. Table 2 lists the measured PV parameters (J_{sc}, V_{oc}, FF, η) for these three samples. Compared to the J_{sc} for the *c*-Si cell (23.7 mA/cm^2), J_{sc} for the PS-1 cell improved to 27.5 mA/cm^2. This improvement indicates an increase in the incident photon flux at the *p*-*n* junction. Figure 4c shows the EQE of these three solar cells. Compared to the *c*-Si cell, the PS-1 cell showed a broad spectral region of photosensitivity to a UV range from 300 nm to 450 nm, suggesting a possible wide optical window due to the PS-1 layer. The overall enhanced EQE for the passivated PS-1 cell is due to the reduction in surface recombination, resulting in a higher J_{sc} of 30.5 mA/cm^2 and FF of 0.550. The η for the PS solar cells with surface passivation increased from 7.93 to 8.54%.

Table 2. PV parameters for PS-1 cell, as-fabricated passivated PS-1 cell, and reference *c*-Si cell.

	c-Si Cell	As-Fabricated PS-1 Cell	Passivated PS-1 Cell
J_{sc} (mA/cm^2) [1]	23.7	27.5	30.5
V_{oc} (V)	0.510	0.510	0.510
FF	0.660	0.300	0.550
η (%) [1]	7.93	4.15	8.54

[1] J_{sc} and η were calculated based on the active area.

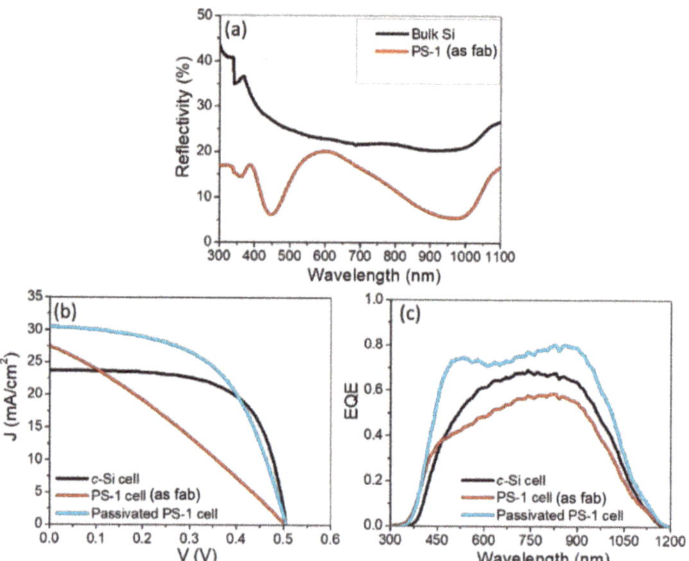

Figure 4. PV characteristics of *c*-Si cell, as-fabricated PS-1 cell and passivated PS-1 cell: (**a**) reflectivity spectra measured for bulk Si and PS-1 layer; (**b**) *J*–*V* curve; (**c**) EQE spectra.

To investigate surface passivation, as shown in the anodization reaction (1), H_2O oxidation kinetics suggest that no chlorine species is incorporated in the oxide during HCl/H_2O oxidation [31], while the addition of HCl to H_2O oxidation ambient could provide uniform oxides with few defects [33]. Moreover, the HCl addition decreases the oxidation rate, which would benefit diffusion control during oxidation of a high aspect-ratio PS layer.

A promising technique for silicon-on-insulator (SOI) technology is oxidized PS at any depth [34]. The bottom-up PS oxidation process is shown schematically in Figure 5a. The PS oxidation rate depends on the density and surface area of the pores of the PS. Due to their differences in density and pore microstructure, the *p*-layer below the *n*-layer is preferentially anodized compared with the *n*-layer Si [31]. The pores of PS can be considered trenches, although anodization reactants travel down from the *n* layer to the *p*-layer. This bottom-up anodization process produces a uniform, broad coverage of oxidation on pore surfaces.

3.3. Electrochemical Passivation Effect in Solar Cells with Thick PS Layer of 191–255 nm

These same etching and passivation processes were used here to fabricate PS-2 layers with thickness of 191–255 nm, and which were then surface-passivated via anodic SiO_2 formation for PS-2 solar cell applications. The uniform morphology of these layers is shown in cross-sectional SEM images in Figure 5b–j. The SiO_2 anodization process was carried out for passivation of the PS-2 layers for different t = 6 s, 10 s, 15 s, 20 s, 30 s, 60 s, 90 s, and 135 s. This scale of SEM images confirmed that this passivation method does not affect the original structure, as evidenced by the uniformity of the porous structure of these passivated samples (Figure 5b–j) that conforms with that of the sample without passivation (Figure 5a).

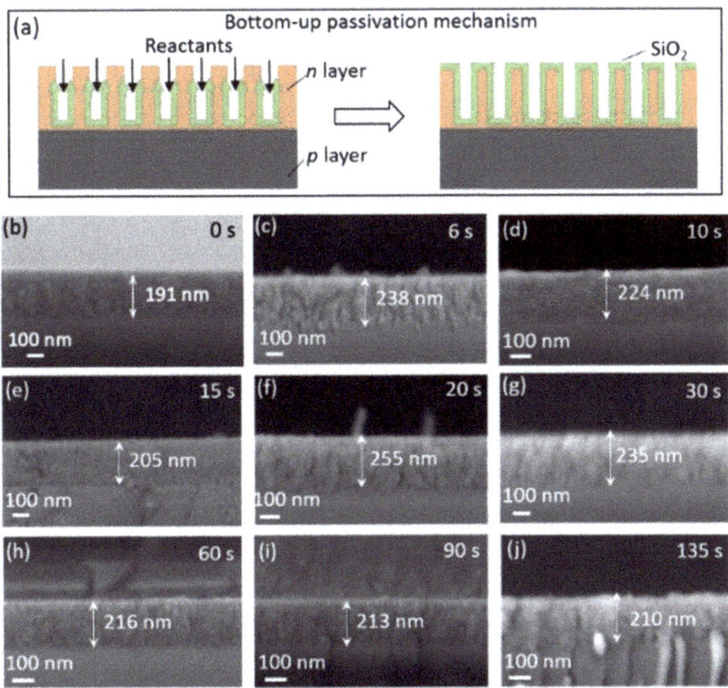

Figure 5. (**a**) Schematic of the bottom-up anodization process of SiO_2 passivated PS-2 layer; (**b**–**j**) cross-sectional SEM images of fabricated PS-2 layers of different thickness (191–255 nm).

Figure 6a shows the *J*–*V* curves under AM1.5 illumination for the PS-2 cell, PS-2-(a-h) cells passivated with anodic SiO_2, and for the reference *c*-Si solar cell. Table 3 lists the measured PV parameters for these samples. All the cells showed good diode-like behavior, although the effect of *t* is striking. The PS-2 cell (i.e., no passivation) achieved an η of 5.15%, whereas the passivated PS-2-e cell (*t* = 30 s) exhibited the highest η of 10.7% (Figure 6b and Table 3). From Figure 6c, although all the cells with a PS layer (data marked with red symbols) show a significant increase in J_{sc} compared to the reference c-Si cell, which indicates improvement due to the anti-reflection property of a PS layer, the passivation effect is not responsible for this improvement in η, as evidenced by J_{sc} being similar in both the PS-2-e cell (with passivation) and the PS-2 cell (no passivation), 29.8 mA/cm^2 and 29.9 mA/cm^2, respectively. The factor that causes this large increase in η is due to a significant increase in *FF* from 0.321 to 0.685. The *t* dependence of *FF* is shown in Figure 6d. For passivated cells (PS-2-(a-h) cells), *FF* increased with increasing *t* up to 30 s and peaked at 0.685. This is attributed to an increase in J_{sc} without significant loss in V_{oc} (Figure 6e). The initial increase in J_{sc} up to *t* of about 30 s indicates a suitable coverage of SiO_2 film on the surfaces of the pores resulting in the best η. These results demonstrate that the bottom-up anodization process successfully fills tiny pores in the PS, and significantly decreases the surface recombination of the PS layer. Longer *t* (>30 s) causes a decrease in J_{sc} due to lower conductivity from SiO_2 caused by overoxidation. As shown in Figure 6e, all the PS-2 solar cells had similar V_{oc}, around 0.5 V, which is lower than the 0.7 V usually obtained from the heterojunction with intrinsic thin layer (HIT) solar cells with a-Si as emitter. This lower V_{oc} might be due to contact resistance losses at the interface between the Si solar cell and the metal contact and to the impact of electrode geometry.

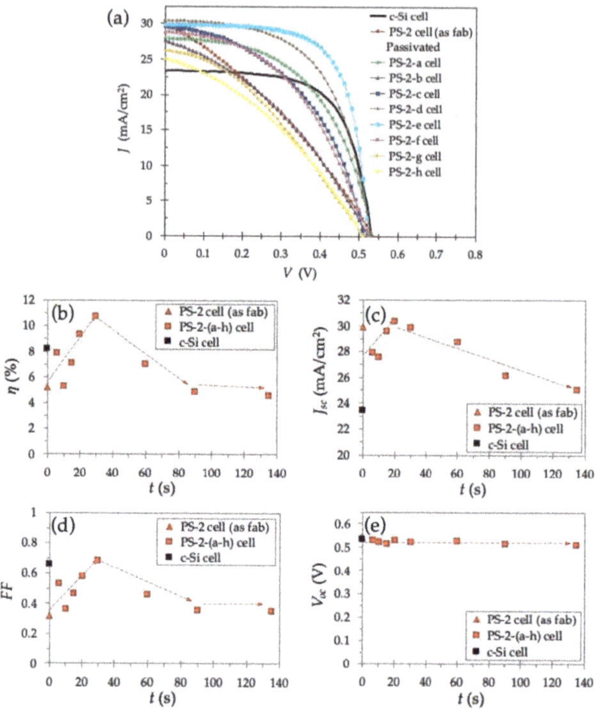

Figure 6. (**a**) *J*–*V* curves; (**b**) η; (**c**) J_{sc}; (**d**) *FF*; (**e**) V_{oc} for as-fabricated PS-2 cell, passivated PS-2-(a-h) cells with different *t*, and reference *c*-Si cell.

Table 3. PV parameters for as-fabricated PS-2 cell, passivated PS-2 cells with different *t*, and reference *c*-Si cell.

Samples	*t* (s)	PS Layer Thickness (nm)	V_{oc} (V)	J_{sc} (mA/cm^2) [1]	*FF*	η (%) [1]
c-Si cell	0	0	0.532	23.4	0.657	8.16
As fabricated PS-2 cell	0	191	0.539	29.9	0.320	5.15
Passivated PS-2-a cell	6	239	0.532	27.9	0.530	7.86
Passivated PS-2-b cell	10	224	0.524	27.6	0.364	5.25
Passivated PS-2-c cell	15	205	0.514	29.6	0.466	7.08
Passivated PS-2-d cell	20	255	0.531	30.4	0.578	9.32
Passivated PS-2-e cell	30	235	0.523	29.8	0.685	10.70
Passivated PS-2-f cell	60	216	0.526	28.7	0.462	6.97
Passivated PS-2-g cell	90	213	0.513	26.2	0.357	4.79
Passivated PS-2-h cell	135	210	0.511	25.0	0.354	4.52

[1] J_{sc} and η were calculated based on the active area.

Figure 7 shows the EQE and IQE for the passivated PS-2 solar cells fabricated for different t compared with an as-fabricated PS-2 solar cell (t = 0 s) and a reference c-Si solar cell. Although the IQE of solar cells is typically calculated from the ratio of its EQE and spectral absorptance as IQE = EQE/(1-reflectance-transmittance), the IQE spectra shown in Figure 7b were calculated by using the reflectivity of the bare surface without a front electrode.

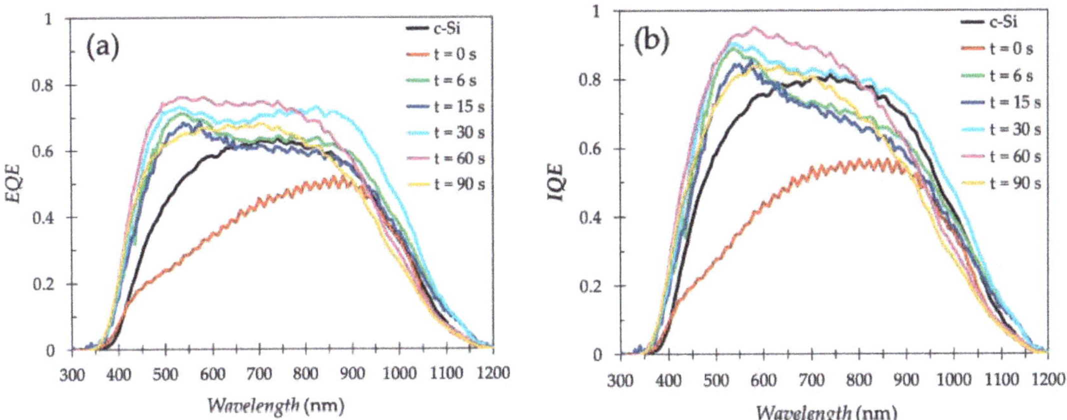

Figure 7. (**a**) EQE; (**b**) IQE of PS-2 cells for different t compared with as-fabricated PS-2 cell and reference c-Si cell.

Throughout most of the spectra at wavelength range of 300–1100 nm, the passivated PS-2 solar cells exhibited enhanced EQE compared to that of the c-Si solar cell, attributed to the PS structure with its unique anti-reflection property. Among all the passivated cells, a notable enhancement in the wavelength range of 700–1100 nm occurred when t = 30 s. This indicates a long diffusion length due to sufficient surface passivation and lower reflectivity at that wavelength range (700–1100 nm), which consequently yields the highest FF. The increase in incident photons trapped at that wavelength range (700–1100 nm) is required for application in thin solar cells because most of the light in the long wavelength region (700–1100 nm) traverses through the semiconductor material unabsorbed. Compared to the passivated PS-2 cells and c-Si cells, the as-fabricated PS-2 cell exhibited the lowest IQE throughout the spectra. This low IQE can be attributed to the front surface recombination at short wavelengths and to the rear recombination. The rear recombination can cause a reduction in the absorption of photons with long wavelengths and low diffusion lengths. The low IQE caused by surface recombination eliminates the advantage of low reflectance. IQE values were significantly improved with the anodic SiO_2 surface passivation (Figure 7b). In particular, a broader spectrum in the wavelength range of 300–1100 nm occurred when t =30 s, indicating a sufficient reduction in surface recombination and improvement in minority carrier diffusion length of the passivated PS-2 cells. Currently, ALD films that are 5, 10, or 20 nm-thick are used for passivation of PS-2 solar cells. However, under AM1.5 illumination, these PS-2 solar cells exhibit unsatisfactory PV parameters, which is attributed to the growth of Al_2O_3 films being diffusion-limited in these high aspect-ratio PS structures [35]. The consequently downward trend in the saturation front process prevents the reactants from penetrating deep along the pore walls. This is further evidence that for PS solar cells, the anodization SiO_2 process is more feasible than that of ALD in surface passivation. In contrast, the dominant dependence of the PS-2 thickness on the performance of such solar cells could not be observed (Figure S1).

Figure 8 shows the R_s (8a) and R_{sh} (8b) of the as-fabricated PS-2 cell (no passivation, t = 0), passivated PS-2-(a-h) solar cells as a function of t, with the c-Si cell as a reference. Compared to the c-Si cell, the as-fabricated PS-2 cell exhibits a relatively high value of

R_s = 18 Ω cm^{-1}, indicating an insufficient movement of light-generated current through the as-fabricated PS-2 layer and the base of the solar cell. More notably, it also corresponds to the lowest R_{sh} of 0.067 kΩ cm^{-1} in the as-fabricated PS-2 cell, due to the power losses in the as-fabricated PS-2 cell caused by the alternate current path to the light-generated current. Because the as-fabricated PS-2 cell did not undergo passivation treatment, a non-uniform oxide layer caused by the native oxide was formed on the PS layer surface. This nonuniformity of the oxide layer acts as a crystal defect that provides an excessive amount of trapping states, which consequently promotes the recombination process. When this recombination current is strong enough, it can act as a shunt [36]. Therefore, despite having a high J_{sc} of 29.9 mA/cm^2 and V_{oc} of 0.539 V similar to the passivated PS-2 cells (Table 3), a significant amount of light-generated current was lost in the PS-2 cell through the high recombination process due to the nonuniformly passivated surface, thus resulting in the low R_{sh}, FF, and η.

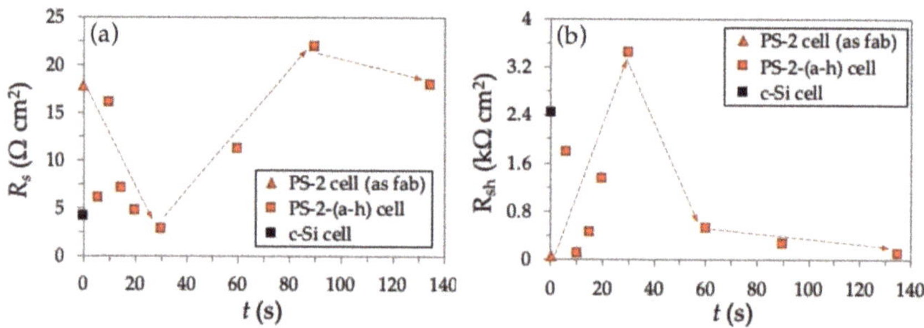

Figure 8. Correlation between t and (**a**) R_s; (**b**) R_{sh} for as-fabricated PS-2 cell, passivated PS-2(a-h) cells, and reference c-Si cell.

For the passivated PS-2-(a-h) solar cells, R_s decreased with increasing t from 0 s to 30 s, and then increased gradually, as shown in Figure 8a. At t = 30 s, the passivated PS-2-e cell reached the lowest R_s of 2.79 Ω cm^{-1}. This is attributed to the enhanced light-generated current through the passivated PS-2-e cell, which had less defect-trapping states as well as less Ag front electrode penetration into the insufficiently passivated pore, and thus leading to a decrease in R_s. However, the additional deposited anodic SiO_2 (at t > 30 s) might turn into an inefficient carrier transport medium. Because the Ag electrode was directly deposited on the passivated PS-2 surface, it might increase the contact resistance between the metal contact and the silicon material. Power losses caused by R_{sh} represent a parallel high-conductivity path across the p-n junction for light-generated current in solar cells. The variation in R_{sh} with t is shown in Figure 8b. The R_{sh} initially increased rapidly, then peaked at 3.45 kΩ cm^{-1} at t = 30 s. This indicates less loss of light-generated current across the p-n junction due to the reduction in recombination by surface passivation of the PS-2 layer. The decrease in R_{sh} with a further increase in t corresponds to the increased tendency of R_s at the same t conditions. At t > 30 s, R_s in the series circuit of solar cells increases due to the overoxidation of anodic SiO_2 deposition between the metal contact and the silicon. Simultaneously, due to an as-yet unknown cause, the recombination rate is increased when the PS layer is passivated for longer than 30 s. A possible mechanism that decreases R_{sh} might originate from the crystal defects resulting from overoxidation [36].

4. Conclusions

In summary, the cross-sectional view of the microstructure of fabricated 146-nm-thick PS-1 and 176-nm-thick PS-2 layers confirmed a lattice matching at the PS/c-Si interface. The lowest reflectivity of approximately 9% was achieved for this thin PS-1 at a wavelength of 700–1100 nm, compared to 18% at 450–700 nm. This highlights the potential of the PS-1 layer as an anti-reflection coating structure in Si solar cells. The broader EQE spectral region of photosensitivity of a passivated PS-1 cell compared to *c*-Si cells at short wavelength suggests a possible wide optical window role of the PS-1 layer.

A bottom-up anodization process successfully produced a uniform SiO_2 passivation layer with high coverage on high aspect-ratio PS-2 layers without disturbing the original structure. A maximum efficiency of 10.7% with the best J–V behavior yielded an FF of 0.685 for the PS-2 cell at an optimized passivation time t of 30 s (PS-2-e cell). Simultaneously, this PS-2 cell had the lowest R_s (2.79 Ω cm^{-1}) and also had the highest R_{sh} (3.45 kΩ cm^{-1}). Such improved PV parameters confirm that the anodic SiO_2 provides a sufficient passivation approach, especially for nanostructuring a PS solar cell. Notable enhancement in EQE in the wavelength range of 700 to 1100 nm improves the light-trapping at long wavelength (700–1100 nm), and thus provides a promising possibility for improving the light absorption in thin-film Si solar cells. Furthermore, the electrochemical anodic oxidation at room temperature can be used as a surface passivation technique for other nanostructured materials because the thickness of the oxidized layer can be precisely controlled by adjusting the combination of current density and treatment period while maintaining the original nanostructure.

Supplementary Materials: The following are available online at https://www.mdpi.com/2079-4991/11/2/459/s1, Figure S1: Dependence of PS thickness on PV parameters of PS-2 cells: (**a**) J_{sc}; (**b**) V_{oc}; (**c**) η; (**d**) FF; (**e**) FF.

Author Contributions: Conceptualization, M.I.; methodology, X.-M.Z., K.M., R.Y., P.S. and M.I.; validation, P.S., X.-M.Z. and M.I.; investigation, P.S., X.-M.Z., R.Y. and A.F.; writing—original draft preparation, X.-M.Z. and P.S.; writing—review and editing, P.S., X.-M.Z., A.F. and M.I.; visualization, P.S. and X.-M.Z.; supervision, M.I. All authors have read and agreed to the published version of the manuscript.

Funding: This research received no external funding.

Data Availability Statement: Data is contained within the article or supplementary material.

Acknowledgments: This work was supported by Center for Low Carbon Society Strategy (LCS), Research and Education Consortium for Innovation of Advanced Integrated Science (CIAiS) and JSPS KAKENHI Grant Number JP16H02451.

Conflicts of Interest: The authors declare no conflict of interest.

References

1. Priolo, F.; Gregorkiewicz, T.; Galli, M.; Krauss, T.F. Silicon nanostructures for photonics and photovoltaics. *Nat. Nanotechnol.* **2014**, *9*, 19–32. [CrossRef]
2. Yoshikawa, K.; Kawasaki, H.; Yoshida, W.; Irie, T.; Konishi, K.; Nakano, K.; Uto, T.; Adachi, D.; Kanematsu, M.; Uzu, H.; et al. Silicon heterojunction solar cell with interdigitated back contacts for a photo over 26%. *Nat. Energy.* **2017**, *2*, 17032. [CrossRef]
3. Shockley, W.; Queisser, H.J. Detailed Balance Limit of Efficiency of p-n Junction Solar Cells. *J. Appl. Phys.* **1961**, *32*, 510–519. [CrossRef]
4. Zhang, X.-M.; Akita, H.; Ihara, M. Epitaxial growth of silicon nanowire arrays at wafer-scale using high-speed rotating-disk CVD for improved light-trapping. *CrystEngComm* **2016**, *18*, 6153. [CrossRef]
5. Wang, L.-X.; Zhou, Z.-Q.; Hao, H.-C.; Lu, M. A porous Si- emitter crystalline-Si solar cell with 18.97% efficiency. *Nanotechnology* **2016**, *27*, 425207. [CrossRef] [PubMed]
6. Lukianov, A.; Ihara, M. Free-standing epitaxial silicon thin films for solar cells grown on double porous layers of silicon and electrochemically oxidized porous silicon dioxide. *Thin Solid Films* **2018**, *648*, 1–7. [CrossRef]
7. Garnett, E.; Yang, P. Light Trapping in Silicon Nanowire Solar Cells. *Nano Lett.* **2010**, *10*, 1082–1087. [CrossRef] [PubMed]
8. Kelzenberg, M.D.; Boettcher, S.W.; Petykiewicz, J.A.; Turner-Evans, D.B.; Putnam, M.C.; Warren, E.L.; Spurgeon, J.M.; Briggs, R.M.; Lewis, N.S.; Atwater, H.A. Enhanced absorption and carrier collection in Si wire arrays for photovoltaic applications. *Nat. Mater.* **2010**, *9*, 239–244. [CrossRef] [PubMed]
9. Dzhafarov, T. Silicon Solar Cells with Nanoporous Silicon Layer. In *Solar Cells—Research and Application Perspectives*; Morales-Acevedo, A., Ed.; IntechOpen: London, UK, 2013; p. 42. ISBN 978-953-51-6313-8.
10. Martí, A.; Araújo, G.L. Limiting efficiencies for photovoltaic energy conversion in multigap systems. *Sol. Energy Mater. Sol. Cells* **1996**, *43*, 203–222. [CrossRef]
11. Chen, H.; Hou, X.; Li, G.; Zhang, F.; Yu, M.; Wang, X. Passivation of porous silicon by wet thermal oxidation. *J. Appl. Phys.* **1996**, *79*, 3282–3285. [CrossRef]

12. Li, G.; Hou, X.; Yuan, S.; Chen, H.; Zhang, F.; Fan, H.; Wang, X. Passivation of light-emitting porous silicon by rapid thermal treatment in NH3. *J. Appl. Phys.* **1996**, *80*, 5967–5970. [CrossRef]
13. Chen, S.; Huang, Y.; Lai, H.; Li, C.; Wang, J. Investigation of passivation of porous silicon at room temperature. *Solid State Commun.* **2007**, *142*, 358–362. [CrossRef]
14. Shiraz, H.G. Effect of anodization time on photovoltaic properties of nanoporous silicon based solar cells. *Sustain. Energy Fuels* **2017**, *1*, 652. [CrossRef]
15. Khezami, L.; Jemai, A.B.; Alhathlool, R.; Rabha, M.B. Electronic quality improvement of crystalline silicon by stain etch-ing-based PS nanostructures for solar cells application. *Sol. Energy* **2016**, *129*, 38. [CrossRef]
16. Aziz, W.J.; Ramizy, A.; Ibrahim, K.; Hassan, Z.; Omar, K. The effect of anti-reflection coating of porous silicon on solar cells efficiency. *Optik* **2011**, *122*, 1462–1465. [CrossRef]
17. Hoex, B.B.; Gielis, J.J.; Van De Sanden, M.R.; Kessels, W.E. On the c-Si surface passivation mechanism by the negative-charge-dielectric Al_2O_3. *J. Appl. Phys.* **2008**, *104*, 113703. [CrossRef]
18. Huang, Z.; Zhong, S.; Hua, X.; Lin, X.; Kong, X.; Dai, N.; Shen, W. An effective way to simultaneous realization of excellent optical and electrical performance in large-scale Si nano/microstructures. *Prog. Photovoltaics: Res. Appl.* **2014**, *23*, 964–972. [CrossRef]
19. Chen, Y.; Zhong, S.; Tan, M.; Shen, W. SiO_2 passivation layer grown by liquid phase deposition for silicon solar cell application. *Front. Energy* **2017**, *11*, 52–59. [CrossRef]
20. Grant, N.E.; Kho, T.; Fong, K.; Franklin, E.; McIntosh, K.; Stocks, M.; Wan, Y.; Wang, E.-C.; Zin, N.; Murphy, J.; et al. Anodic oxidations: Excellent process durability and surface passivation for high efficiency silicon solar cells. *Sol. Energy Mater. Sol. Cells* **2019**, *203*, 110155. [CrossRef]
21. Oh, J.; Yuan, H.-C.; Branz, H.M. An 18.2%-efficient black-silicon solar cell achieved through control of carrier recombination in nanostructures. *Nat. Nanotechnol.* **2012**, *7*, 743–748. [CrossRef]
22. Patil, J.J.; Smith, B.D.; Grossman, J.C. Ultra-high aspect ratio functional nanoporous silicon via nucleated catalysts. *RSC Adv.* **2017**, *7*, 11537–11542. [CrossRef]
23. Ben Slama, S.; Hajji, M.; Ezzaouia, H. Crystallization of amorphous silicon thin films deposited by PECVD on nickel-metalized porous silicon. *Nanoscale Res. Lett.* **2012**, *7*, 464. [CrossRef]
24. Porous Silicon Homepage. Available online: https://www.poroussilicon.com/wafer-etching.html (accessed on 18 January 2021).
25. Gautier, G.; Defforge, T.; Desplobain, S.; Billoué, J.; Capelle, M.; Povéda, P.; Vanga, K.; Lu, B.; Bardet, B.; Lascaud, J.; et al. Porous silicon in microelectronics: From academic studies to industry. *ECS Trans.* **2015**, *69*, 123. [CrossRef]
26. Boehringer, M.; Artmann, H.; Witt, K. Porous Silicon in a Semiconductor Manufacturing Environment. *J. Microelectromechanical Syst.* **2012**, *21*, 1375–1381. [CrossRef]
27. Hasegawa, K.; Takazawa, C.; Fujita, M.; Noda, S.; Ihara, M. Critical effect of nanometer-size surface roughness of a porous Si seed layer on the defect density of epitaxial Si films for solar cells by rapid vapor deposition. *CrystEngComm* **2018**, *20*, 1774–1778. [CrossRef]
28. Shibahara, R.; Hasegawa, K.; Fave, A.; Fourmond, E.; Noda, S.; Ihara, M. *Electrical Properties of Monocrystalline Thin Film Si for Solar Cells Fabricated By Rapid Vapor Deposition with Nano-Surface Controlling Double Layer Porous Si in H_2*; ECS Meeting Abstracts; IOP Publishing: Bristol, UK, 2019. [CrossRef]
29. Lukianov, A.; Murakami, K.; Takazawa, C.; Ihara, M. Formation of the seed layers for layer-transfer process silicon solar cells by zone-heating recrystallization of porous silicon structures. *Appl. Phys. Lett.* **2016**, *108*, 213904. [CrossRef]
30. Bruggeman, D. Calculation of different physical constants of heterogen substances: Dielectric constants and conductibility of mixtures from isotrop substances. *Ann. Phys.* **1935**, *416*, 665–679. [CrossRef]
31. Kriegler, R.J. *Semiconductor Silicon*; Huff, H.R., Burgess, R.R., Eds.; Electrochemical Society: Princeton, NJ, USA, 1973; p. 363.
32. Beale, M.; Benjamin, J.; Uren, M.; Chew, N.; Cullis, A. An experimental and theoretical study of the formation and microstructure of porous silicon. *J. Cryst. Growth* **1985**, *73*, 622–636. [CrossRef]
33. Deal, B.E. Thermal Oxidation Kinetics of Silicon in Pyrogenic H_2O and 5% HCI/H_2O Mixtures. *J. Electrochem. Soc.* **1978**, *125*, 576. [CrossRef]
34. Patridge, S.L. Silicon-on-insulator technology. *IEE Proc. E* **1986**, *133*, 66.
35. Knoops, H.C.M.; Potts, S.E.; Bol, A.A.; Kessels, W.M.M. Atomic Layer Deposition. *Handb. Cryst. Growth* **2015**, *3*, 1101–1134.
36. Breitenstein, O.; Rakotoniaina, J.P.; Al Rifai, M.H.; Werner, M. Shunt types in crystalline silicon solar cells. *Prog. Photovolt. Res. Appl.* **2004**, *12*, 529–538. [CrossRef]

Article

Concept for Efficient Light Harvesting in Perovskite Materials via Solar Harvester with Multi-Functional Folded Electrode

Mao-Qugn Wei [1], Yu-Sheng Lai [2,*], Po-Hsien Tseng [1], Mei-Yi Li [2], Cheng-Ming Huang [2] and Fu-Hsiang Ko [1,*]

[1] Department of Materials Science and Engineering, National Chiao Tung University, Hsinchu 300093, Taiwan; barry720919@gmail.com (M.-Q.W.); a19890302@gmail.com (P.-H.T.)
[2] Taiwan Semiconductor Research Institute, National Applied Research Laboratories, Hsinchu 300091, Taiwan; meiyi@narlabs.org.tw (M.-Y.L.); pfhaung@narlabs.org.tw (C.-M.H.)
* Correspondence: yslai@narlabs.org.tw (Y.-S.L.); fhko@mail.nctu.edu.tw (F.-H.K.); Tel.: +886-03-577-3693-7532 (Y.-S.L.); +886-03-571-2121-55803 (F.-H.K.)

Abstract: Conventional electrodes in typical photodetectors only conduct electrical signals and introduce high optical reflection, impacting the optical-to-electrical conversion efficiency. The created surface solar harvester with a multi-functional folded electrode (MFFE) realizes both a three-dimensional Schottky junction with a larger light detecting area as well as low optical reflection from 300 nm (ultra-violet light) to 1100 nm (near-infrared light) broadly without an additional anti-reflection layer. The MFFE needs silicon etching following the lithography process. The metal silver was deposited over structured silicon, completing the whole device simply. According to the experimental results, the width ratio of the bottom side to the top side in MFFE was 15.75, and it showed an optical reflection of 5–7% within the major solar spectrum of AM1.5G by the gradient refractive index effect and the multi-scattering phenomenon simultaneously. While the perovskite materials were deposited over the MFFE structure of the solar harvester, the three-dimensional electrode with lower optical reflection benefitted the perovskite solar cell with a larger detecting area and an additional anti-reflection function to absorb solar energy more efficiently. In this concept, because of the thin stacked film in the perovskite solar cell, the solar energy could be harvested by the prepared Schottky junction of the solar harvester again, except for the optical absorption of the perovskite materials. Moreover, the perovskite materials deposited over the MFFE structure could not absorb near-infrared (NIR) energies to become transparent. The NIR light could be harvested by the light detecting junction of the solar harvester to generate effective photocurrent output additionally for extending the detection capability of perovskite solar cell further. In this work, the concept of integration of a conventional perovskite solar cell with a silicon-based solar harvester having an MFFE structure was proposed and is expected to harvest broadband light energies under low optical reflection and enhance the solar energy conversion efficiency.

Keywords: broadband light harvesting; three-dimensional Schottky junction; perovskite solar cell; photoresponsivity

Citation: Wei, M.-Q.; Lai, Y.-S.; Tseng, P.-H.; Li, M.-Y.; Huang, C.-M.; Ko, F.-H. Concept for Efficient Light Harvesting in Perovskite Materials via Solar Harvester with Multi-Functional Folded Electrode. *Nanomaterials* **2021**, *11*, 3362. https://doi.org/10.3390/nano11123362

Academic Editors: Aurora Rizzo and Byungwoo Park

Received: 30 September 2021
Accepted: 8 December 2021
Published: 11 December 2021

1. Introduction

Electricity generation has been provided mainly by fossil fuels (like natural gas, oil, nuclear energy, and coal), leading to the global warming issues. Solar energy is a renewable, clean, and environmental fuel to be a suitable candidate for generating green energy. Because the optical behaviors include the reflection, transmission, and absorption as the incident radiation illuminates the light energy harvester, the incident light energy harvested must have as high optical absorption as possible under the conditions of low optical reflection over the surface of the energy harvester for better optical energy-to-electricity conversion efficiency. Furthermore, the effective separation of photo-generation carriers in the light detecting region provides the photocurrent or photovoltage output usefully. It is well-known that nanostructure engineering can increase incidence light harvesting to improve conversion efficiency [1–13]. In 2013, Chong Liu et al. [4] combined TiO_2 nanowire and p-type Si for solar to hydrogen (STH) applications. In their study, TiO_2 nanowire and p-type Si had different energy band gaps to achieve the broadband optical operating range. Furthermore, although the TiO_2 nanowire structure can increase incidence light harvesting to improve the power conversion efficiency, the efficiency of 0.1% is not high enough for commercial photovoltaic application. Moreover, the previous reports used metal nanoparticles [8,9] or core–shell nanostructures [5,6,10] with localized surface plasmon resonance (LSPR) effects through changing the dimension of nanoparticles (gold or silver) to obtain higher optical absorption in specific light wavelengths. These optical behaviors are strongly dependent on the shape and the size of nanoparticles conducting the narrower optical absorption regime in the solar spectrum to limit the optical-to-electrical conversion capability. The gold nanorod structure with surface plasmonic effect was provided by Mubeen et al. to achieve an STH efficiency of 0.1%. Although Mubeen et al. created the Schottky barrier at the gold nanorod/TiO_2 interface to separate electron–hole pairs, the higher barrier still impacted the conversion efficiency [12]. Consequently, integrating LSPR with nanostructures or nanoparticles operates in the narrower solar spectrum without obvious broadband light harvesting, resulting in poorer power conversion efficiency.

In recent reports [14–29], solar energy harvest can be perovskite solar cells for higher conversion efficiency. The perovskite solar cell always absorbs the light wavelength from 400 nm to 800 nm because of the energy band gap of materials for preparing the perovskite-based solar cell. Many researchers improved the efficiency by adjusting the materials directly to conduct two kinds of perovskite solar cells, namely two terminal (2T; single junction) [15–24] and four terminal (4T, stacked cells) [25–29]. In 1959, H. Kallmann (1959) [14] demonstrated the first organic solar cell (anthracene single crystals), which has an efficiency of 2×10^{-6}%. The perovskite structure with metal-halide (ABX_3) can enhance the performance, introducing the reliability issue because of the non-stable material. Several works made the effort of material stability through energy band gap adjustment of ABX_3. For example, Michael M. Lee. et al. [20,21] proposed a perovskite solar cell with wide band gap energy of 1.55 eV by $CH_3NH_3PbI_2Cl$ to achieve an efficiency of 10.9% (400–800 nm). Eric. T. Hoke et al. [22] and David P. McMeekin et al. [23] also presented the higher energy band gap, from 1.85 eV to 1.68 eV, and that of 1.74 eV, respectively, to bring the non-stable issue into the design. For overcoming the disadvantages of structure stability, photo-carriers transmission, and the wide band gap energy (the narrower optical regime) of materials in 2T perovskite solar cells, the four-terminal (4T) perovskite solar cell was provided. Colin D. Bailie et al. [24] provided the 4T perovskite solar cell composed of silver nanowire with Spiro-OMeTAD (for the hole-transporting layer), mesoporous titanium dioxide (for the electron-transporting layer), and the stacked tandem configuration onto copper indium gallium diselenide (CIGS). UV light illuminates TiO_2 because oxygen can be generated to impact the band gap of the perovskite material, bringing the limited conversion efficiency of 13.7%. Although the ETL composed of SnO_2 [25] or PCBM/PEIE [26] (without the oxygen's influence under UV regime) could achieve efficiencies of 18.4% and 19.2%, respectively, broadband optical energy harvesting is still limited by the energy band gap of the material itself. Furthermore, once the broadband light conversion can be implemented, the conversion efficiency should be improved further. For this purpose, combining a hetero-junction structure or CIGS with a perovskite solar cell provides the possibility to extend the optical absorption capability. These methodologies always conduct complex fabrication in whole device integration.

In this work, a surface solar energy harvester with a multi-functional folded electrode (MFFE) was prepared for three-dimensional light harvesting (Schottky junction) with broadband optical absorption (300–1100 nm) and effective photocurrent conduction simultaneously at the same device structure. The solar harvester could be fabricated through the conventional semiconductor process without altering any process step. After completing the whole harvester with the MFFE structure, optical anti-reflection and light harvesting were realized in the single MFFE structure simultaneously. Typically, the metal electrode with high optical reflection behavior impacts optical-to-electrical conversion capability. The higher optical reflection behavior is usually from the intense change in the optical property of n (the refractive index) along the incident light transition path. Moreover, the anti-reflection structure over flat metal film is too complex to provide the broadband low optical reflection. Besides, the trade-off between the area of the metal electrode and the light detecting region is another change, because the smaller metal electrode brings the longest diffusion distance of photo-generated carriers with a higher recombination possibility. The surface solar harvester is created by integrating the conventional Schottky junction with the structured silicon surface in a specific three-dimensional periodical optical structure to form the MFFE structure.

The MFFE structure provides the gradient refractive index for lowering the intense refractive index change in the interface between air and metal to achieve the low optical reflection within the solar spectrum (AM1.5G) as broadly as possible. For easy incident light harvesting, the thin metal is deposited over the MFFE structure, forming the Schottky junction to harvest the incident light. It is worth mentioning that the correlation between incident optical intensity and light penetration depth is exponential decay, indicating that the light detecting junction must be located near the Si surface for the better light harvesting capability. Because of the designed Schottky junction in the MFFE structure just below the metal electrode, experimental results also indicate the high photoresponsivity and low optical reflection as the incident light wavelength from 300 nm to 1100 nm, compared with the contrast device without any optical structure. At the same time, theoretical analysis is also provided to confirm the MFFE structure with an obvious gradient refractive index phenomenon. The significant breakthrough can be expected to break the trade-off between the area of the metal electrode and the light detection region to provide excellent solar harvesting and carrier collection within the same MFFE structure. Without an extra anti-reflection structure, the silicon width ratio of 15.75 in the MFFE structure reveals the lowest optical reflection of 3–5% from the ultra-violet to the near-infrared regime broadly. Furthermore, this work also realizes 2.5 times and 5 times the external quantum efficiency (EQE) and the responsivity of the contrast device without an optical structure, respectively. As the perovskite solar cell is deposited over the three-dimensional MFFE optical structure, the former possesses more detecting area and additional anti-reflection function in the visible regime to enhance the conversion efficiency of the perovskite solar cell further. More, it is due to the thin perovskite materials that the visible light energy can be absorbed by the solar harvester and the perovskite solar cell at the same time. The responsivity of 0.5 A/W in the created harvester peaks around 900 nm, benefiting the NIR light harvesting even though the perovskite cell cannot absorb NIR light. Consequently, the concept of integration of the perovskite solar cell with the surface solar harvester profits broadband energy harvesting efficiently.

2. Materials, Fabrication, and Methods

2.1. Device Fabrication

The mainstream complementary metal-oxide-semiconductor (CMOS) technology was used to fabricate the solar harvester without altering any standard processes. Fabrication began with the pattern formation of the three-dimensional structure in the surface of silicon through electron-beam lithography (EBL, Leica Weprint 200 e-beam, Leica Weprint 200 e-beam, Leica Camera AG Corp., Wetzlar, Germany). The dry etching system (TCP-9400, Lam Research TCP 9400 Poly Etcher, Lam Research Corp., Fremont, CA, USA) provided the silicon removal process according to the lithography pattern to achieve the trench depth of 1.2–1.3 μm. After stripping off the photo resister from the surface of the silicon, the three-dimensional structures were prepared with the diameter and the period of 0.4 μm and 0.8 μm, respectively. Then, for reducing the traps (dangling bound) induced from the silicon etching process, the silicon dioxide of 20 nm was formed and removed by 49% of the hydrofluoric acid (HF and DI water with a volume ration of 1:50) at 25 °C immediately to repair the damaged silicon surface. Using sputter and physical vapor deposition (FSE-Cluster-PVD, F.S.E Corp., New Taipei City, Taiwan), the silver film (topside electrode) and TiN film (backside electrode) with thicknesses of 20 nm and 50 nm, respectively, were obtained. As the silver deposited over these optical structures, the structured metal was fabricated in the same process simultaneously to achieve process simplification in our device. Due to the differences between the work function of metal electrode and the fermi-level of n-type silicon, the Schottky junction and the ohmic junction could be provided in the interfaces between silver and n-type silicon and that among n-type silicon and TiN, respectively. Furthermore, the created three-dimensional Schottky junction in the solar harvester achieved a larger light harvesting area in the fixed device size and the photoelectron injection due to the band diagram of the Schottky junction. Obviously, the fabrication of the proposed device with MFFE structure was as straightforward as fabricating the conventional Schottky diode. Because of the diameter and the period of 0.4 μm and 0.8 μm in the designed optical structures, respectively, the pattern definition process could use i-line lithography for lower fabrication cost and high throughput instead of the expensive electron-beam lithography process in the future. Figure S1 describes the brief flow to illustrate the design concept feasibility of integrating the solar harvester with the perovskite solar cell in the future. Indium tin oxide (ITO) is usually the bottom electrode of the conventional perovskite solar cell. The perovskite materials can be coated following the ITO being deposited over the three-dimensional silver film in the MEEF structure of the solar harvester. The interface between the silver (top electrode of the solar harvester) and ITO electrode in the perovskite solar cell should be a good ohmic contact without any issues of fabrication. It also illustrates the good electrical connection to confirm that the perovskite solar cell and the solar harvester are in the series connection electrically shown in Figure S1.

2.2. Metrology, Optical Simulation, and Device Characterization

The prepared optical structures were imaged using transmission electron microscopy (TEM, JEM-2010F, JEOL Corp., Tokyo, Japan) for confirming the device structure, especially the width ratio of the bottom silicon to the top one in the MFFE structure. The three-dimensional finite-difference time-domain (3D-FDTD, Ansys/LUMERICAL Corp., Vancouver, Canada) method was employed to simulate the optical behaviors (optical reflection), the distribution of the electrical field in the proposed device, and those in the contrast device lacking the optical structure. In the device verification, an optical spectrometer (Hitachi, U4100, Hitach Corp., Tokyo, Japan) was used to investigate the optical reflection of all prepared samples. Consisting of the monochromatic having the light source of the Xe lamp, the optical filter (300–1100 nm) and the source-measurement-unit (Keithley 2400, Keithley Corp., Solon, OH, USA) provided the optical-to-electrical conversion efficiency data such as photoresponsivity and the current–voltage characteristics in this work.

3. Results and Discussion

3.1. Device Design

Figure 1a illustrates the light energy distribution from 400 nm to 1100 nm within the solar spectrum of AM1.5G. In this regime, the silicon-based energy harvester could absorb all the incident energies because of the silicon energy band gap of 1.12 eV only. In contrast to the silicon material, the perovskite solar cell with the larger energy band gap always absorbed the light wavelength from 400 nm to 800 nm. Within the solar spectrum of AM1.5G in Figure 1a, only 72% the incident solar energy could be harvested by the perovskite solar cell to contribute to the optical-to-electrical conversion. The energy band gap of material utilized in the perovskite solar cell was too large to absorb the near-infrared (NIR) energies, introducing the waste of incident energies (28% the intensity spectrum in AM1.5G). For harvesting the additional NIR energies, the conventional perovskite solar cell could be integrated with the silicon-based energy harvester. While the perovskite solar cell was stacked over the MFFE structure of the solar harvester, the NIR energies could still reach the light detecting area of the Schottky junction of the solar harvester easily. In this study, the concept of integration of a perovskite solar cell with a silicon-based optical energy harvester may be the candidate for not only the optical energy absorption as broadly as possible but also the enhancement of optical-to-electrical conversion capability further. In device fabrication, the perovskite solar cell can be fabricated through multi-thin film deposition with total film thickness of a few hundred nm. These film depositions can be easy to reproduce over the MFFE structure for the purpose of perovskite solar cells stacked over the solar harvester. For electrical consideration, the bottom electrode in the perovskite solar cell and the top metal of the MFFE are ITO and silver, respectively. It was clearly confirmed that there was a good electrical connection, as the perovskite solar cell and the solar harvester were connected in series electrically. In optical energy harvesting, because of the three-dimensional MFFE structure in the desired solar harvester, the deposited perovskite materials can achieve a larger dimension for light harvesting compared with that of a conventional perovskite solar cell. Furthermore, the perovskite solar cell only absorbs the light energies from 400 nm to 800 nm. The lower optical reflection in the visible spectrum is crucial to the perovskite solar cell. The suggested solar harvester with the MFFE structure realizes the broadband low optical reflection, benefiting the visible light absorption of perovskite materials as long as the perovskite solar cell is stacked over the solar harvester. Because of the thin stacked perovskite materials over the MFFE structure of the solar harvester, the incident light energies can be harvested by the solar harvester again to provide more photocurrent output relative to the conventional perovskite solar cell. Moreover, the near-infrared energies can pass through the perovskite solar cell to arrive at the silicon-based solar harvester for generating electron–hole pairs additionally in the interface between n-type silicon and silver (MFFE region). Therefore, comparing the single perovskite solar cell with that stacked over the silicon-based solar harvester, the latter can be expected to realize light harvesting broadly for providing higher optical-to-electrical conversion efficiency within solar spectrum of AM1.5G potentially.

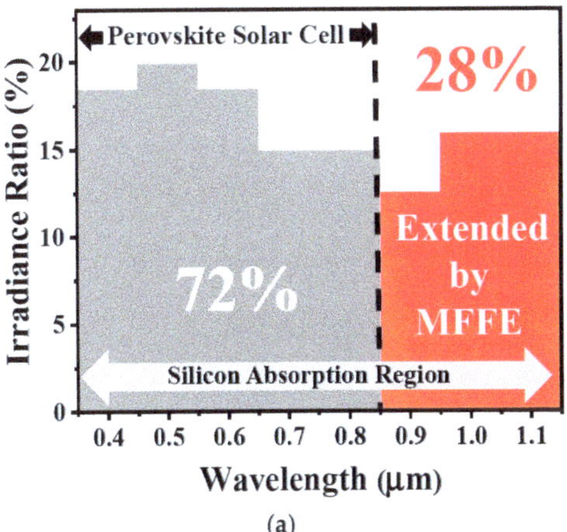

(a)

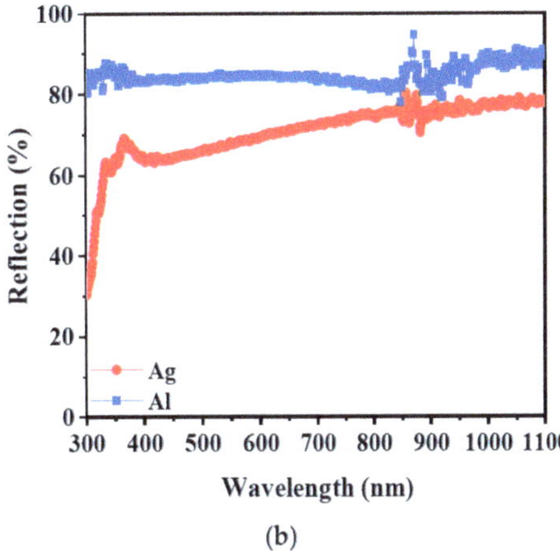

(b)

Figure 1. (**a**) The irradiation ratio distribution within AM1.5G. The perovskite solar cell absorbs only 72% of the irradiation spectrum. Combining the multi-functional folded electrode (MFFE) with the perovskite solar cell benefits the harvest of additional near-infrared light energy for enhancing the conversion efficiency further; (**b**) the optical reflection of metal silver (Ag) and that of aluminum (Al). Significantly, the high optical reflection can be observed without optical structure, and silver has the inter-band transition phenomenon (low optical reflection) in the ultra-violet regime.

Generally, the conventional metal electrode of the silicon-based optical energy harvester with high optical reflection naturally impacts the optical absorption. The anti-reflection structure over the metal electrode is difficult to conduct the complex device fabrication. Figure 1b displays the optical reflection behavior of metal aluminum (Al) and that of metal silver (Ag) in the ultra-violet (UV, 300–400 nm), the visible (400–700 nm), and the NIR (700–1100 nm) regime. Significantly, the optical reflection of Al without any anti-reflection structures is as high as 80% at least. Under the same verification condition, in the case of Ag, it shows the lower optical reflection in the UV regime and in the visible regime compared with metal Al, which is due to the difference of optical properties (n and k) in Al and Ag and the inter-band transition phenomenon of silver in the UV regime additionally. We believe the MFFE with silver will achieve high UV light harvesting capability relative to that with metal of Al. The Schottky junction can be created in the interface between silver metal and n-type silicon, having the depletion region in silicon to harvest light energy. Significantly, the light harvesting regime is near the silicon surface, benefiting absorption of the incident light power. Typically, the incident energy shows exponential decay accompanying the increase of light penetration depth into the silicon. The shorter the light penetration depth in silicon is, the higher the incident intensity can be obtained for harvesting and converting by the surface solar energy harvester. Consequently, the Schottky junction could be the candidate for the surface optical harvester, profiting to harvest light with the advantages of easy fabrication and a large junction area in the three-dimensional MFFE structure for harvesting incident light energy simultaneously. The prepared devices in this study are illustrated in Figure 2a–c, which are the cross-sectional SEM images to illustrate the different optical structures, namely the multi-functional folded electrode (MFFE), Type I, and Type II, respectively. The formation of these structures in Figure 2 is silicon etching following the lithography process for transferring the optical structure patterns. Then, the metal silver was sputtered to complete the whole device. This process prevents the manufacture of the optical structures over the surface of metal electrode directly and benefits the large (three-dimensional) Schottky junction area formation with the folded metal structure simply. All these device design concepts can be easily achieved through one lithography process, one silicon etching, and metal deposition with a metal thickness of around 15 nm.

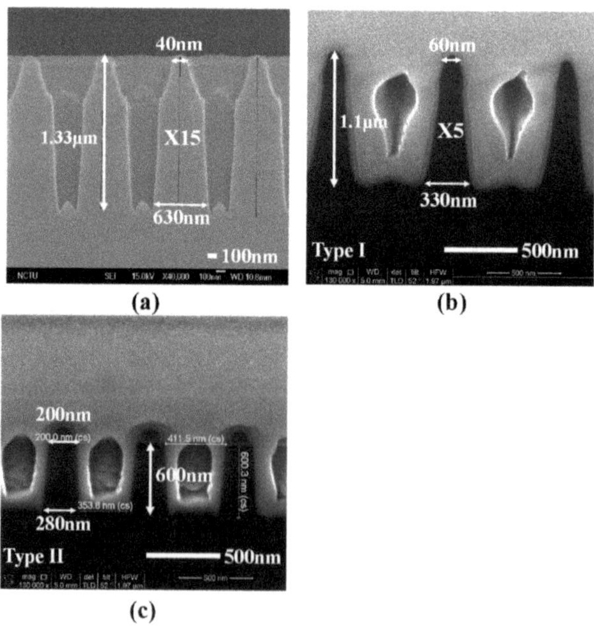

Figure 2. The cross-sectional view of prepared optical structures. The multi-functional folded electrode (MFFE), Type I, and Type II are shown in (**a**–**c**), respectively ("X15" and "X5" marked in (**a**,**b**), respectively, are the width ratio of the bottom silicon to the top one, indicating the different tilt in the prepared structures).

3.2. Theoretical Analysis of Effective Refractive Index

In Figure 1b, the metal without any optical structure shows the high optical reflection phenomenon. Herein, the theoretical analysis will be provided to confirm the suggested MFFE with lower optical reflection within the solar spectrum of AM1.5G. The low optical reflection can be realized as long as the change of refractive index is insignificant along the light traveling path. For example, the refractive index of air is quite different from a metal surface conducting a significant optical reflection. The results of theoretical analysis are illustrated in Figure 3 for confirming the existence of a gradient refractive effect in the prepared structures (Figure 2) to realize the lower and broader optical reflection behavior in this work. The verifying results of light wavelengths are UV light, visible light, and NIR light, as displayed in Figure 3a–c, respectively. The analysis calculated the effective optical property n (the refractive index) contributed by the n of silicon and that of air with the variant ratio of silicon to air accompanied with the different incident light penetration depth. In Figure 3a, the approximate linear change in effective refractive index n indicates the optical structure, MFFE, with the gradient refractive index effect that is beneficial to optical reflection reduction. Generally, high optical reflection occurs in the interface, having a dramatic change of refractive index, such as pure metal in air. While the silicon is etched like a tapered shape in the MFFE with the period and the top silicon width of 0.8 μm and 0.04 μm, respectively, this structure provides the effective refractive index produced by the silicon and air under some light penetration depth and multi-scattering phenomenon simultaneously. In the same way, under deeper light penetration depth, it can provide another effective refractive index to form the gradient refractive index effect, lowering the significant difference in refractive index between silicon and air. For this purpose, the MFFE has the inclined plane in Figure 2a to provide the gradient ratio of silicon to air for stimulating an effective gradient refractive index effect and the multi-scattering optically. According to the calculated results in Figure 3, the MFFE structure achieved a greater linear change in refractive index as light penetration depth increased compared with the other optical structures, namely Type I and Type II. Significantly, the effective refractive index of Type II presented an obvious sudden change no matter what the incident lights were. As the incident light met the prepared structure of Type II, it was due to the widest top silicon width of 0.2 μm in Figure 2c, introducing the additional optical reflection and optical interference to impact the gradient refractive index effect. Additionally, although two optical structures, MFFE and Type I, had similar etched depths of silicon and the pitch of optical structure arrangement, the appearance diversity of the optical structure introduced completely different optical behavior. Structure "Type I" had a top silicon width of 0.06 μm, which was wider than that of MFFE (~40 nm), and a narrower bottom silicon width of 0.33 μm compared with that of MFFE (~0.63 μm). Investigating the "effective n" of MFFE in Figure 3c still realizes a more linear gradient refractive index change compared with the contrast structure, Type I and Type II. In the longer light wavelength like the NIR regime, the dimension of the optical structure MFFE is fixed, and it will conduct the unexpected optical behavior to impact the gradient refractive index effect slightly. Through the calculated results of effective refractive index in Figure 3a–c, these results imply the proposed MFFE with the capability of broadband optical reflection reduction without an external anti-reflection structure. The simulation results through the finite-difference time-domain (3D-FDTD) software provide evidence for the optical reflection reduction phenomenon in the prepared MFFE structure illustrated in Figure 3d. In Figure 3d, the proposed surface solar energy harvester with MFFE realized the lower optical reflection within the solar spectrum broadly compared with the contrast device without any optical structure. In the inset of Figure 3d, the brightness area in the simulated result indicated the strongest electrical field region (optical absorption). It is clear to investigate the electrical field confined in the suggested MFFE for re-confirming the multi-scattering phenomenon and the incident light harvested by MFFE to demonstrate lower optical reflection.

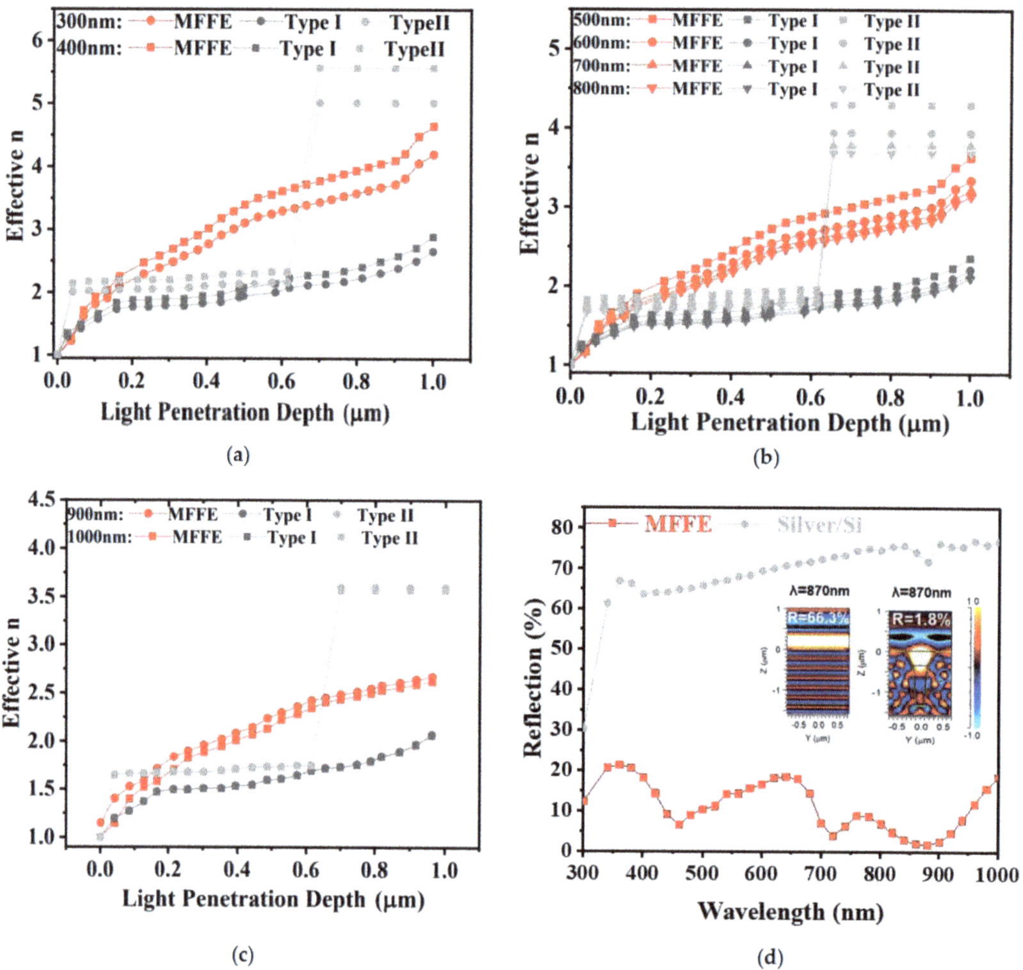

Figure 3. The calculated correlation between effective refractive index (n) and light penetration depth in (**a**) ultra-violet light; (**b**) visible light; and (**c**) near-infrared light. These results illustrate that MFFE provides the linear gradient refractive index compared with the others in this work; (**d**) the simulated optical reflection by three-dimensional finite-differential time-domain (FDTD) software.

3.3. Verification of Optical Behaviors

According to the results shown in Figures 2a and 3a–c, the MFFE device can be expected to provide a significant gradient ratio of silicon to air for stimulating the effective gradient refractive index effect to lower the significance of the optical reflection theoretically. The verified results of optical reflection are displayed in Figure 4. The lack of any anti-reflection structure or gradient refractive index effect in the silver/Si sample introduces the highest optical reflection of 60–70% from UV to NIR regimes. It also points out the low optical-to-electrical conversion capability in the silver/Si sample at the same time. The device with the structure of Type II having the dramatic change in effective refractive index (Figure 3) also conducts obvious optical interference and the optical reflection as high as 20% because of the widest top width of silicon and the shallower (~0.6 μm) depth in the designed structure of Type II compared with that of MFFE (~1.3 μm). According to Figure 2c, the top and bottom silicon width of the Type II structure were 0.2 μm and 0.28 μm, respectively;

the width difference between the top silicon and bottom one was too similar in the optical structure to provide the efficient gradient ratio of silicon to air for stimulating the effective gradient refractive index effect. Without any anti-reflection structure, the measured results in the inset of Figure 4 illustrate that the significant low optical reflection (<7%) within broadband incident light wavelengths from 300 nm to 1100 nm can be achieved in the proposed solar harvester with the MFFE structure relative to all the prepared samples. Two optical structures of MFFE and Type I have similar etched depth of silicon and the pitch of optical structure arrangement in the SEM images of Figure 2a,b, respectively. Comparing the reflection of the device with MFFE and that with Type I shown in the inset of Figure 4, the former provided the highest optical reflection of only 7% with insignificant optical interference in the visible to NIR regime successfully. It confirms the results of theoretical analysis about the gradient refractive index (in Figure 3a–c) and the multi-scattering phenomenon (electrical field confined phenomenon in Figure 3d) in the MFFE structure. Moreover, the width ratio of the bottom silicon to the top one was 5.5 (bottom/top: 0.33 μm/0.06 μm) in the Type I structure, introducing a poorer gradient spacing width of the optical structure without stimulating a more linear gradient refractive index effect compared with that of 15.75 (bottom/top: 0.63 μm/0.04 μm) in the MFFE structure. Therefore, devices with Type I structures display higher optical reflection and optical interference, which is clearly shown in the insert of Figure 4.

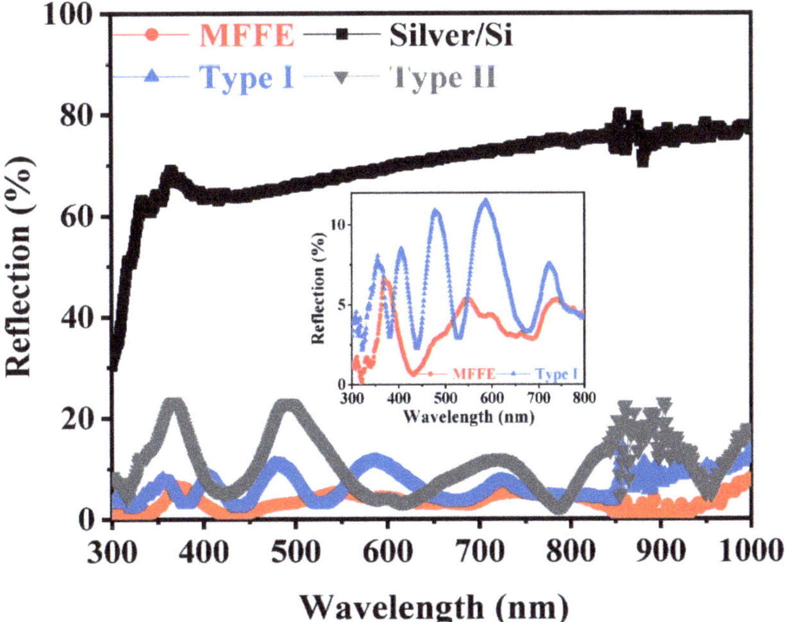

Figure 4. The measured optical reflection through the optical integrating sphere system. The MFFE provides the lowest optical reflection from incident light wavelength of 300 nm to that of 1000 nm. The optical reflection can be as low as 5–7%, illustrated in the inset of Figure 4 to confirm the broadband light harvesting in MFFE without anti-reflection structure over silver additionally.

3.4. Device Characterization

In Figure 5a, the current–voltage characteristics provide evidence about the existence of the Schottky junction between Al and n-type silicon and that between Ag and n-type silicon in the same MFFE structure. As devices are in dark environments, the cut-in voltage of devices with Al deposited is smaller than that of devices with Ag, indicating the lower Schottky barrier height in the Al case. It may be the obvious Schottky junction barrier lowering effect because of more interface defects in the metallurgical interface. Moreover, more photocurrent output can be provided in the device with the silver-deposited MFFE structure (relative to the device with the Al-deposited MFFE structure) under the same illuminating intensity. According to these results, the metal silver was chosen to form the suggested solar energy harvester in this work. The external quantum efficiency (EQE) and the responsivity are illustrated in Figure 5b,c, respectively, to investigate the optical-to-electrical conversion capabilities in the prepared samples. Clearly, as long as the device with the MFFE structure, and the highest EQE and responsivity can be achieved, confirming the desired gradient refractive index effect, the optical multi-absorption, and the electrical field confined phenomenon in the MFFE structure simultaneously. Moreover, in Figure 5b,c, the sample named "silver/Si" is the contrast device without any optical structure having the highest optical reflection shown in Figure 4. Because of the thinner thickness of silver in this work, the incident light could penetrate through the metal to arrive at the depletion region of the Schottky junction between silver and n-type silicon. Generally, the optical property, extinction coefficient k, is higher in the UV regime than that in the visible regime. As the shorter incident light wavelength illuminates the contrast device, it conducts the significant performance of EQE and that of responsivity, implying the optical absorption behavior is dominated by silicon itself directly. According to the measured results in Figure 5b,c, the device named "silver/Si" is still unable to provide the highest EQE or responsivity like the device with MFFE, confirming that the suggested MFFE structure is crucial to light harvesting capability instead of high UV light absorption in silicon itself. Unlike the operation principle of contrast devices, the MFFE structure provides the extra optical anti-reflection effect to harvest more incident light energy as broadly as possible within the solar spectrum of AM1.5G for enhancing the EQE or responsivity further. According to responsivity data in Figure 5c, the spectrum peaked at around 900 nm due to the MFFE structure with the dimensional arrangement in the diameter (0.4 μm) and period (0.8 μm). Based on the verified result, the design concepts of multi-functional electrodes were demonstrated successfully, including the three-dimensional surface optical energy harvester, broadband optical anti-reflection from UV to NIR regime without additional optical structure over metal silver, and the simple fabrication for preparing the device as straight forwardly as the conventional Schottky diode simultaneously. Moreover, Figure 5d illustrates the design concept feasibility of integration the perovskite solar cell with the suggested solar harvester with the MFFE structure. The process details are shown in Figure S1. Through the verified results of the solar harvester with the MFFE, the created solar harvester had the peak photo-responsivity of around 900 nm within the AM1.5G spectrum in this work. The process of perovskite solar cell deposition over the MFFE structure is easy to be realized by thin film coating. That also provides a greater detecting region in a fixed device size because of the three-dimensional MFFE optical structure. Further, in conventional perovskite solar cells, the optical reflection reduction in the visible light regime is crucial to improve the energy conversion efficiency. The prepared MFFE structure can provide the additional optical anti-reflection functionality to benefit the visible light energies absorbed by the perovskite solar cell, except the optical absorption of the Schottky junction in the solar harvester. Figure 5d illustrates the concept of this study in combining a conventional perovskite solar cell with a silicon-based solar harvester having a three-dimensional MFFE structure. The solar harvester realizes the broadband low optical reflection benefiting the visible light absorption of perovskite materials as the perovskite solar cell is stacked over the solar harvester. Because of the thin perovskite materials, the incident light energies can be harvested by the solar harvester again to provide the enhanced energy conversion efficiency compared with the conventional perovskite solar cell. In the NIR regime, the perovskite solar cell becomes transparent, which is due to the larger band gap of perovskite materials than that in silicon. The NIR energies can pass through the thin perovskite material to be absorbed by the Schottky junction in the solar harvester. In this work, the concept of integration of a conventional perovskite solar cell with a silicon-based solar harvester with an MFFE structure has proposed and expected to harvest broadband light energies under low optical reflection as well as the enhancement of solar energy conversion efficiency. Table 1 illustrates the device performance of the solar harvester relative to that of the reported perovskite solar cell in optical absorption capability (device operation range), external quantum efficiency (EQE), and photocurrent density simultaneously.

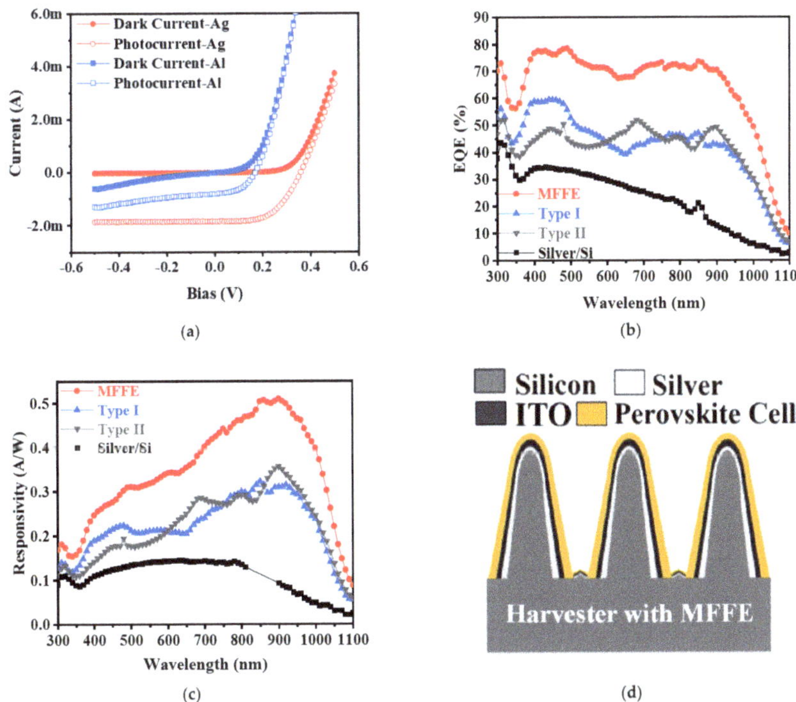

Figure 5. (**a**) Current–voltage (I–V) characteristics of the MFFE device with deposition of different metal materials (Ag and Al); (**b**) the external quantum efficiency (EQE) spectral response; (**c**) the responsivity of the MFFE structure compared to other structures (type I, type II, and silver/si structure); (**d**) concept of integrating a perovskite solar cell with a prepared solar harvester with MFFE.

Table 1. Comparisons of device performances.

	Science [23]	Energy Environ. Sci. [25]	Energy Environ. Sci. [29]	This Work
Optical Absorption	400–700 nm	400–700 nm	400–700 nm	300–1100 nm
Light Harvest Regime	Multi-junction	Multi-junction	Multi-junction	Single-junction
EQE (%) (400–700 nm)	70% @ 400 nm 90% @ 500–700 nm	~5% @ 400 nm 70% @ 500–700nm	80% @ 400 nm 70% @ 600 nm 60% @ 700 nm	70% @ 300 nm 80% @ 400 nm 70% @ 500–700 nm
EQE (%) (800–1000 nm)	N/A	N/A	N/A	70% @ 800–900 nm 50% @ 1000 nm 10% @ 1100 nm
Current Density (mA/cm^2)	19.2	17.2	17.5	2

4. Conclusions

In this work, a surface solar harvester with an MFFE for efficient harvesting of the light energies from 300 nm (UV) to 1100 nm (NIR) was proposed and verified by a three-dimensional Schottky junction, revealing the broadband optical reflection of 5–7% (UV to NIR regime) and photogeneration carrier's collection/conduction at the same structure to demonstrate the intension of a multi-functional electrode successfully. Without altering any mainstream semiconductor process, the MFFE structure provides the gradient reflective index and multi-scattering phenomenon to reduce the optical reflection broadly for enhancing the optical harvesting capability within the solar spectrum of AM1.5G. According to the experimental results, the ratio of bottom silicon width to that of the top in the MFFE is 15.75, illustrating the lowest optical reflection. This is in agreement with the theoretical analysis of the gradient reflective index effect. It is worth mentioning that the solar harvester with the MFFE lacking an extra anti-reflection structure still achieves 2.5 times and 5 times the EQE and the responsivity of the contrast device without any optical structure, respectively. Moreover, the spectral response of the solar harvester peaks near 900 nm because of the diameter of 0.4 μm and period of 0.8 μm in the MFFE structure. As the perovskite solar cell is deposited over the three-dimensional MFFE optical structure, the former possesses more detecting area and the additional anti-reflection function to enhance the conversion efficiency further. The responsivity of 0.5 A/W in the created harvester peaks around 900 nm, benefiting from the NIR light harvesting in the depletion region of MFFE even though the perovskite cell cannot absorb NIR light. Consequently, the concept of integration of a perovskite solar cell with a surface solar harvester was proposed and is expected to harvest broad light energies (by a perovskite solar cell and a Schottky junction in a solar harvester) under low optical reflection (by an MEEF structure) for enhancing solar energy conversion efficiency further.

Supplementary Materials: The following are available online at https://www.mdpi.com/article/10.3390/nano11123362/s1, Figure S1: The brief flow for illustrating the design concept in this work about integrating solar harvester with the perovskite solar cell.

Author Contributions: M.-Q.W., Y.-S.L. and F.-H.K. conceived the idea and designed the experiments. P.-H.T. and C.-M.H. contributed to the FDTD simulations; M.-Y.L. prepared the samples and contributed to photocurrent measurement. M.-Q.W., Y.-S.L. and F.-H.K. wrote the paper. All authors contributed to data analysis and scientific discussion. All authors have read and agreed to the published version of the manuscript.

Funding: This research was funded by the Ministry of Science and Technology, Taiwan, under the contracts MOST 107-2221-E-492-022-MY3 and MOST 110-2221-E-492-004.

Data Availability Statement: This study did not report any data.

Acknowledgments: The authors would like to thank all colleagues and students who contributed to this study. We also wish to thank the Taiwan Semiconductor Research Center, Taiwan, and National Synchrotron Radiation Research Center, Taiwan, for the device fabrication support and the material verification, respectively.

Conflicts of Interest: The authors declare no conflict of interest.

References

1. Kelzenberg, M.D.; Boettcher, S.W.; Petykiewicz, J.A.; Turner-Evans, D.B.; Putnam, M.C.; Warren, E.L.; Spurgeon, J.M.; Briggs, R.M.; Lewis, N.S.; Atwater, H.A. Enhanced absorption and carrier collection in Si wire arrays for photovoltaic applications. *Nat. Mater.* **2010**, *9*, 239. [CrossRef] [PubMed]
2. Zhu, J.; Hsu, C.-M.; Yu, Z.; Fan, S.; Cui, Y. Nanodome solar cells with efficient light management and self-cleaning. *Nano Lett.* **2009**, *10*, 1979–1984. [CrossRef] [PubMed]
3. Zhu, J.; Yu, Z.; Burkhard, G.F.; Hsu, C.-M.; Connor, S.T.; Xu, Y.; Wang, Q.; McGehee, M.; Fan, S.; Cui, Y. Optical absorption enhancement in amorphous silicon nanowire and nanocone arrays. *Nano Lett.* **2008**, *9*, 279–282. [CrossRef] [PubMed]
4. Agarwal, D.; Aspetti, C.O.; Cargnello, M.; Ren, M.; Yoo, J.; Murray, C.B.; Agarwal, R. Engineering localized surface plasmon interactions in gold by silicon nanowire for enhanced heating and photocatalysis. *Nano Lett.* **2017**, *17*, 1839–1845. [CrossRef] [PubMed]

5. Liu, C.; Tang, J.; Chen, H.M.; Liu, B.; Yang, P. A fully integrated nanosystem of semiconductor nanowires for direct solar water splitting. *Nano Lett.* **2013**, *13*, 2989–2992. [CrossRef]
6. Zhou, L.; Yu, X.; Zhu, J. Metal-core/semiconductor-shell nanocones for broadband solar absorption enhancement. *Nano Lett.* **2014**, *14*, 1093–1098. [CrossRef]
7. Aydin, K.; Ferry, V.E.; Briggs, R.M.; Atwater, H.A. Broadband polarization-independent resonant light absorption using ultrathin plasmonic super absorbers. *Nat. Commun.* **2011**, *2*, 517. [CrossRef]
8. Baffou, G.; Quidant, R.; García de Abajo, F.J. Nanoscale control of optical heating in complex plasmonic systems. *ACS Nano* **2010**, *4*, 709–716. [CrossRef]
9. Govorov, A.O.; Zhang, W.; Skeini, T.; Richardson, H.; Lee, J.; Kotov, N.A. Gold nanoparticle ensembles as heaters and actuators: Melting and collective plasmon resonances. *Nanoscale Res. Lett.* **2006**, *1*, 84. [CrossRef]
10. Yu, X.; Bi, J.; Yang, G.; Tao, H.; Yang, S. Synergistic effect induced high photothermal performance of Au nanorod@ Cu7S4 yolk–shell nanooctahedron particles. *J. Phys. Chem. C* **2016**, *120*, 24533–24541. [CrossRef]
11. Mubeen, S.; Lee, J.; Singh, N.; Krämer, S.; Stucky, G.D.; Moskovits, M. An autonomous photosynthetic device in which all charge carriers derive from surface plasmons. *Nat. Nanotechnol.* **2013**, *8*, 247. [CrossRef]
12. Lee, J.; Mubeen, S.; Ji, X.; Stucky, G.D.; Moskovits, M. Plasmonic photoanodes for solar water splitting with visible light. *Nano Lett.* **2012**, *12*, 5014–5019. [CrossRef]
13. Hua, B.; Lin, Q.; Zhang, Q.; Fan, Z. Efficient photon management with nanostructures for photovoltaics. *Nanoscale* **2013**, *5*, 6627–6640. [CrossRef]
14. Kallmann, H.; Pope, M. Photovoltaic Effect in Organic Crystals. *J. Chem. Phys.* **1959**, *30*, 585–586. [CrossRef]
15. Todorov, T.; Gershon, T.; Gunawan, O.; Sturdevant, C.; Guha, S. Perovskite-kesterite monolithic tandem solar cells with high open-circuit voltage. *Appl. Phys. Lett.* **2014**, *105*, 173902. [CrossRef]
16. Kojima, A.; Teshima, K.; Shirai, Y.; Miyasaka, T. Organometal halide perovskites as visible-light sensitizers for photovoltaic cells. *J. Am. Chem. Soc.* **2009**, *131*, 6050–6051. [CrossRef]
17. Eperon, G.E.; Stranks, S.D.; Menelaou, C.; Johnston, M.B.; Herz, L.M.; Snaith, H.J. Formamidinium lead trihalide: A broadly tunable perovskite for efficient planar heterojunction solar cells. *Energy Environ. Sci.* **2014**, *7*, 982–988. [CrossRef]
18. Noh, J.H.; Im, S.H.; Heo, J.H.; Mandal, T.N.; Seok, S.I. Chemical management for colorful, efficient, and stable inorganic–organic hybrid nanostructured solar cells. *Nano Lett.* **2013**, *13*, 1764–1769. [CrossRef] [PubMed]
19. Anaya, M.; Lozano, G.; Calvo, M.E.; Míguez, H. ABX3 perovskites for tandem solar cells. *Joule* **2017**, *1*, 769–793. [CrossRef]
20. Lee, M.M.; Teuscher, J.; Miyasaka, T.; Murakami, T.N.; Snaith, H.J. Efficient hybrid solar cells based on meso-superstructured organometal halide perovskites. *Science* **2012**, *338*, 643–647. [CrossRef]
21. Snaith, H.J. Perovskites: The emergence of a new era for low-cost, high-efficiency solar cells. *J. Phys. Chem. Lett.* **2013**, *4*, 3623–3630. [CrossRef]
22. Hoke, E.T.; Slotcavage, D.J.; Dohner, E.R.; Bowring, A.R.; Karunadasa, H.I.; McGehee, M.D. Reversible photo-induced trap formation in mixed-halide hybrid perovskites for photovoltaics. *Chem. Sci.* **2015**, *6*, 613–617. [CrossRef]
23. McMeekin, D.P.; Sadoughi, G.; Rehman, W.; Eperon, G.E.; Saliba, M.; Hörantner, M.T.; Haghighirad, A.; Sakai, N.; Korte, L.; Rech, B. A mixed-cation lead mixed-halide perovskite absorber for tandem solar cells. *Science* **2016**, *351*, 151–155. [CrossRef]
24. Da, Y.; Xuan, Y.; Li, Q. Quantifying energy losses in planar perovskite solar cells. *Sol. Energy Mater. Sol. Cells* **2018**, *174*, 206–213. [CrossRef]
25. Albrecht, S.; Saliba, M.; Baena, J.P.C.; Lang, F.; Kegelmann, L.; Mews, M.; Steier, L.; Abate, A.; Rappich, J.; Korte, L. Monolithic perovskite/silicon-heterojunction tandem solar cells processed at low temperature. *Energy Environ. Sci.* **2016**, *9*, 81–88. [CrossRef]
26. Werner, J.; Weng, C.-H.; Walter, A.; Fesquet, L.; Seif, J.P.; De Wolf, S.; Niesen, B.; Ballif, C. Efficient monolithic perovskite/silicon tandem solar cell with cell area > 1 cm^2. *J. Phys. Chem. Lett.* **2016**, *7*, 161–166. [CrossRef] [PubMed]
27. Werner, J.; Barraud, L.; Walter, A.; Bräuninger, M.; Sahli, F.; Sacchetto, D.; Tétreault, N.; Paviet-Salomon, B.; Moon, S.-J.; Allebeé, C. Efficient near-infrared-transparent perovskite solar cells enabling direct comparison of 4-terminal and monolithic perovskite/silicon tandem cells. *ACS Energy Lett.* **2016**, *1*, 474–480. [CrossRef]
28. Bush, K.A.; Palmstrom, A.F.; Zhengshan, J.Y.; Boccard, M.; Cheacharoen, R.; Mailoa, J.P.; McMeekin, D.P.; Hoye, R.L.; Bailie, C.D.; Leijtens, T. 23.6%-efficient monolithic perovskite/silicon tandem solar cells with improved stability. *Nat. Energy* **2017**, *2*, 17009. [CrossRef]
29. Bailie, C.D.; Christoforo, M.G.; Mailoa, J.P.; Bowring, A.R.; Unger, E.L.; Nguyen, W.H.; Burschka, J.; Pellet, N.; Lee, J.Z.; Grätzel, M. Semi-transparent perovskite solar cells for tandems with silicon and CIGS. *Energy Environ. Sci.* **2015**, *8*, 956–963. [CrossRef]

Communication

Wavelength-Conversion-Material-Mediated Semiconductor Wafer Bonding for Smart Optoelectronic Interconnects

Kodai Kishibe, Soichiro Hirata, Ryoichi Inoue, Tatsushi Yamashita and Katsuaki Tanabe *

Department of Chemical Engineering, Kyoto University, Kyoto 615-8510, Japan;
k.kishibe@cheme.kyoto-u.ac.jp (K.K.); s.hirata@cheme.kyoto-u.ac.jp (S.H.); r.inoue.ku@gmail.com (R.I.); t.yamashita@cheme.kyoto-u.ac.jp (T.Y.)
* Correspondence: tanabe@cheme.kyoto-u.ac.jp

Received: 2 November 2019; Accepted: 5 December 2019; Published: 6 December 2019

Abstract: A new concept of semiconductor wafer bonding, mediated by optical wavelength conversion materials, is proposed and demonstrated. The fabrication scheme provides simultaneous bond formation and interfacial function generation, leading to efficient device production. Wavelength-converting functionalized semiconductor interfacial engineering is realized by utilizing an adhesive viscous organic matrix with embedded fluorescent particles. The bonding is carried out in ambient air at room temperature and therefore provides a cost advantage with regard to device manufacturing. Distinct wavelength conversion, from ultraviolet into visible, and high mechanical stabilities and electrical conductivities in the bonded interfaces are verified, demonstrating their versatility for practical applications. This bonding and interfacial scheme can improve the performance and structural flexibility of optoelectronic devices, such as solar cells, by allowing the spectral light incidence suitable for each photovoltaic material, and photonic integrated circuits, by delivering the respective preferred frequencies to the optical amplifier, modulator, waveguide, and detector materials.

Keywords: semiconductor; interface; wafer bonding; frequency conversion; optoelectronics; photonic device; solar cell; photonic integrated circuit

1. Introduction

The wafer bonding technique [1–3] is used to generate semiconductor heterostructures with low defect densities, which are difficult to obtain by the conventional epitaxial growth methods, owing to crystalline lattice mismatches. Therefore, semiconductor wafer bonding is promising for the realization of high-performance optoelectronic components and has been employed to fabricate various heterostructured devices, such as lasers [4–7], light-emitting diodes [8], photodetectors [9], and solar cells [6,10,11]. Here, we present optical wavelength-converting material (WCM)-mediated wafer bonding, as a means to simultaneously provide bond formation and interfacial function generation. This study is part of a series of developments regarding functional bonding interfaces; we have previously developed graphene-mediated semiconductor wafer bonding [12]. There exist various semiconductor wafer bonding schemes, such as semiconductor-to-semiconductor direct bonding [6,10,11] and oxide- [4,5], metal- [8], and polymer-mediated bonding [7]. However, there has been no WCM-mediated bonding reported so far, to the best of our knowledge. Our novel semiconductor-bonding scheme may improve, for example, multijunction photovoltaic cell efficiencies, by converting the wavelength of the light transmitted through the upper subcell to one that is highly absorbed by the lower subcell [13,14], as conceptually depicted in Figure 1. It also improves the performance of hybrid optical transceivers in photonic integrated circuits [15–18] by delivering suitable frequencies to each optical amplifier, modulator, waveguide, and detector material. Furthermore, our bonding is performed at room temperature and does not require heating, unlike the conventional wafer direct bonding processes, and therefore exhibits no risk of damaging the active material or retarding the production line.

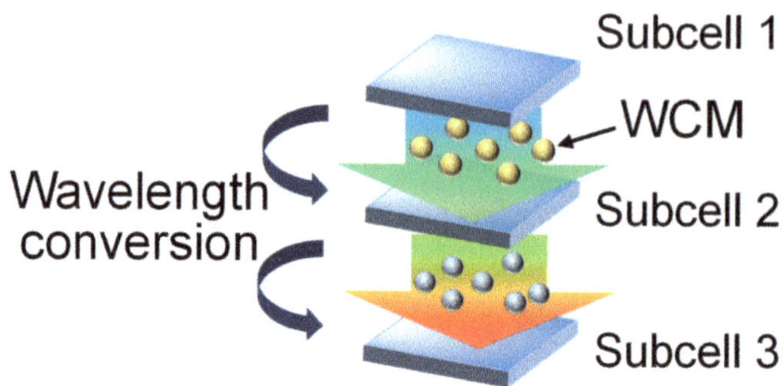

Figure 1. Conceptual drawing of the application of the wavelength-converting material (WCM)-mediated bonding for multijunction solar cells.

2. Experimental Methods

The polished surfaces of boron-doped *p*-type <100> Si wafers, with a doping concentration of ~1×10^{19} cm^{-3} and a thickness of 280 μm, were first coated using photoresist for the purpose of protecting the Si surfaces to be bonded in the dicing process into 64 mm^2. The diced wafers were dipped in acetone for 5 min to remove the photoresist coating and to degrease the Si surfaces to be bonded. They were then submerged in 9% hydrofluoric acid for 1 min to remove the SiO_2 native oxide layer formed on the Si wafers. We used a commercially available WCM of 4,7-bis(4-*tert*-butylphenyl)-2-octylbenzotriazole (RAYCREA, Nitto Denko Corp., Osaka, Japan) [19]—a fluorescent dye compound. However, the bare WCM was composed of solid-state particles and therefore could hardly be incorporated stably, in its original form, in semiconductor wafer bonding interfaces. Therefore, we used a hydrogel material as an adhesive and a viscous organic matrix to embed the fluorescent particles. A 2.5 w/v% polyacrylamide (PAM) aqueous solution (aq.) was prepared by mixing PAM powder with deionized water and stirring well to prevent the aggregation of the adhesive PAM particles. WCM, which was ground with a mortar to obtain uniform particle diameters, was mixed with the 2.5 w/v% PAM aq., and five types of mixtures (0, 1×10^{-4}, 1×10^{-3}, 5×10^{-3}, and 1×10^{-2} g/mL) were prepared. The prepared hydrogel containing WCM was then uniformly spin-coated onto the Si wafer. For a higher reproducibility of the experimental results, this spin-coating process was repeated three times. The Si piece coated with the hydrogel containing WCM was bonded to a bare Si piece at room temperature under a uniaxial pressure of 0.1 MPaG. Incidentally, the use of a WCM that can be molecularly dispersed in a polymer binder would have been technically more desirable, but such a WCM was not commercially available, and we were not equipped with the ability to synthesize it. Therefore, we employed the particulate WCM in this work.

The normal detachment stresses were measured as bonded interfacial strengths for the bonded samples. We connected the outer surface of the bonded sample to a digital spring weight scaler via a solid wire firmly attached to the sample surface using a household adhesive glue. Then, we pulled the scaler outward in the direction normal to the sample die until the bonded sample was debonded while the weight scaler recorded the maximum force at the point of delamination. For electrical measurements, an Au–Ge–Ni alloy (80:10:10 wt %) and pure Au layers with thicknesses of 30 and 150 nm, respectively, were sequentially deposited on the outer sides of the bonded samples as ohmic electrodes via thermal evaporation. For the interfacial observation, we used an infrared transmission analyzer (IRise-T, Moritex Corp., Asaka, Japan) with an optical wavelength of 1.2 μm. For the optical measurements of WCM, a sample containing 0.03 g/cm^2 of WCM between glass plates was prepared in the same manner as above but with no surface pretreatment. In this study, for the sake of simplicity, we used Si wafers as the representative semiconductor material, but our concept can be easily extended to other semiconductors, as numerous wafer bonding experimental demonstrations have been reported, thus far, between dissimilar semiconductor materials [4–11].

3. Results and Discussion

Figure 2 shows a cross-sectional scanning electron microscope image of the bonded interface (WCM concentration: 1×10^{-2} g/mL). As observed in this image, the wafers are firmly and uniformly in contact with each other, with sufficient mechanical strength to endure the cleavage of the bonded sample. The WCM-containing interlayer in this example may seem relatively thick, considering the field of thin-film optoelectronic devices. According to the required degree of the wavelength-converting function and interfacial mechanical stability, the interlayer thickness can be controlled in each application. As a general design direction, a thicker interlayer provides a sufficient wavelength conversion efficiency and higher bondability by mitigating the semiconductor surface roughness, whereas a thinner interlayer supports the advantages in the device weight, production cost, and throughput. The WCM uniformity in the interlayer may be improved by increasing the dispersion of the WCM particles, as discussed later. We measured the dependence of the bonding strength on the WCM concentration; the results are shown in Figure 3. At the present stage, the experimental reproducibility of the bonding strength is relatively low, and we therefore plotted the best measured values for each bonding condition. As observed in these results, the majority of the samples exhibited sufficient bonded interfacial strength to endure a series of optoelectronic device fabrication processes and user operations. The hydrogen bonds stemming from PAM presumably cause the adhesion to the semiconductor surfaces [20,21]. More specifically, hydrogen bonds may form between the $-NH_2$ groups of PAM and the Si surface terminated by –OH groups, owing to the water contained in PAM [20]. In addition, the PAM matrix holds the WCM particles, suppressing their sedimentation. This is due to its viscosity, induced by the entanglement of PAM polymer chains. The results shown in Figure 3 indicate that, as the concentration of WCM increases, the interfacial bonding strength decreases. To further analyze the cause of this result, we observed the inner structure of the sample (WCM concentration: 1×10^{-2} g/mL) using an infrared transmission analyzer (Figure 4a). The black spots in Figure 4a identify the WCM via the accompanying emission image, and this image verifies that the WCM particles are not dispersed uniformly; rather, they appear aggregated. This random dispersion of WCM is due to our having no technical control of the spatial distribution of WCM at the present stage and also, presumably, to the energetic stability in the aggregated state of the WCM particles. Such clustered WCM particles may cause intense roughness that the PAM cannot mitigate. Therefore, the interfacial bonding strength significantly decreases at higher concentrations of WCM, as we observed larger aggregates for higher WCM concentrations (Figure 4). In addition, the wafers may experience serious damage, as evidenced by the cracks observed in Figure 4. This applies especially to higher WCM concentrations, because the bonding pressure is concentrated in particular areas of the agglomerated particles. Therefore, to improve the properties of our WCM-containing interfaces—such as the mechanical strength, electrical conductivity, and optical transparency—it may be effective to increase the dispersion of WCM by selecting the optimum preparation process conditions, such as the organic solvent species, the spin-coating rotation velocity, and the coating repetition. To account for the WCM concentration vs. the interfacial stability trade-off observed in Figure 3, it is important to adopt the optimal organic species and thickness of the matrix and the dispersion status of the WCM particles, according to the property requirements for each application. For instance, a design principle can be described as follows: a relatively thick interlayer with low-concentration WCM can fulfill the demands of both high mechanical strength and high optical conversion efficiency.

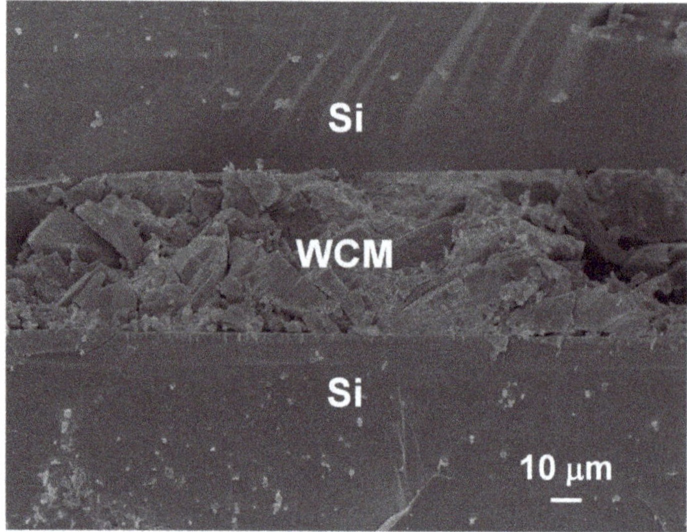

Figure 2. Cross-sectional scanning electron microscope image of a bonded interface (WCM concentration: 1×10^{-2} g/mL).

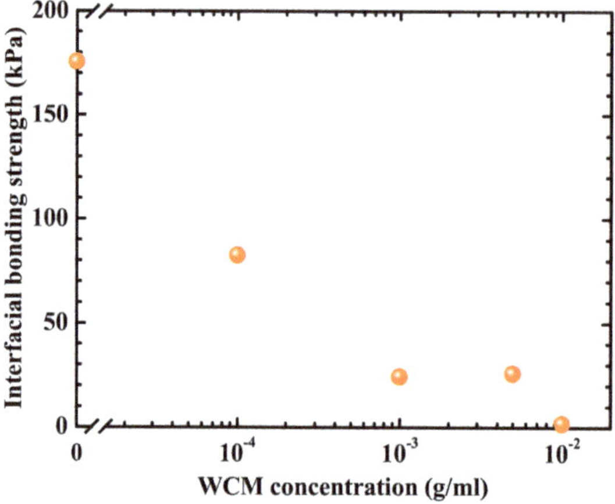

Figure 3. Interfacial bonding strength vs. WCM concentration.

Figure 5 shows the current–voltage (*I*–*V*) characteristics of the bonded interfaces for various concentrations of WCM. An ohmic interfacial electrical conductance with relatively low resistivity is observed when bonding with lower WCM concentrations. PAM, which is a hydrophilic polymer material, can contain water internally, and the water-originated ions are known to act as electrical carriers and provide electric conductance [21]. Intuitively, one might think that the presence of water has a negative effect on electronic component fabrication. However, in hydrogels, the water molecules are stably embedded in the matrices of organic polymer chains, and hydrogel-based devices have been successfully fabricated and operated [21], as in the field of organic electronics. As the concentration of WCM increases, the conductivity decreases, presumably because the highly agglomerated WCM for high WCM concentration conditions can induce voids between the wafers, as indicated in Figure 4. However, the strong decrease of the electrical conductivity with the increased concentration of WCM cannot be explained simply by a decreased width of the conductive joint, because the observed decrease is much larger than the loss of the joint area represented by Figure 4. The bonding interlayer thicknesses for the samples with WCM concentrations of 0 and 1×10^{-2} g/mL were measured at about 5 and 50 μm, respectively, by scanning electron microscopy. The interlayer thickness increase with WCM concentration can be attributed to the generation of larger WCM aggregates for higher WCM concentrations, as observed in Figure 4. Consequently, such large, and therefore tall, aggregated WCM clusters may mechanically resist the interlayer compression through the applied pressure in the bonding process. In this way, the bonding interlayer thickness becomes significantly larger for higher WCM concentrations. Overall, the unintentional combinative effect of such a decrease in the area and an increase in the length of the electrically conductive channel may have resulted, in our experiments, in the significant decrease of the interfacial electrical conductivity with WCM concentration. An increase in dispersion may be effective in obtaining, simultaneously, a high WCM concentration and a high electrical conductivity. A similar conclusion was reached in the previous case of WCM concentration vs. mechanical stability investigation, although the thickened dilute WCM interlayer strategy may also be applicable in this case. In addition, WCM particles with sizes much smaller than the total interlayer thickness may improve the density–conductivity trade-off. Moreover, the doping of electroconductive polymers, such as polyacetylene or polythiophene [22], in PAM may work effectively to produce a higher electrical conductivity.

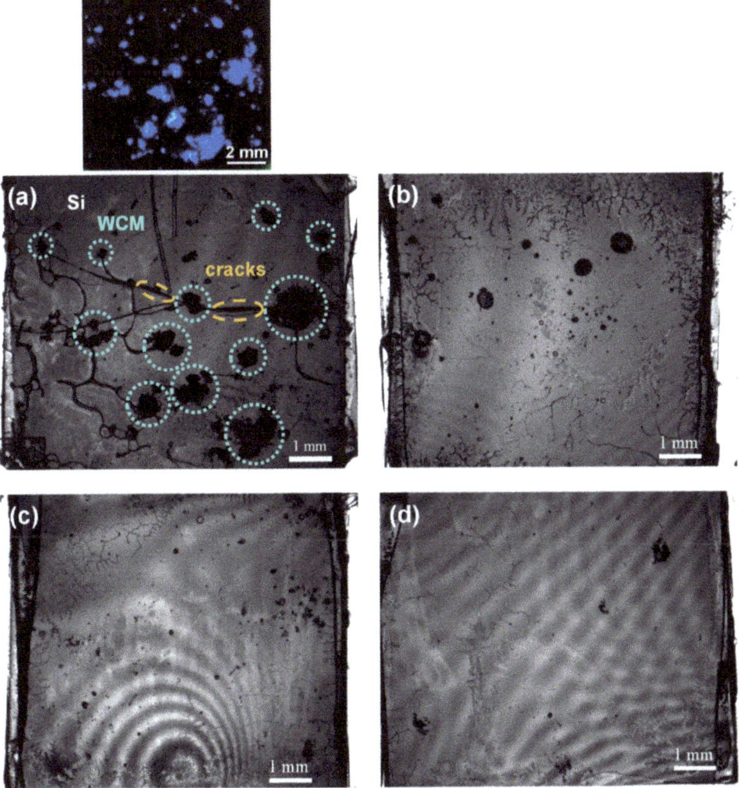

Figure 4. Infrared inner images of bonded interfaces (WCM concentration: (**a**) 1×10^{-2} g/mL, (**b**) 5×10^{-3} g/mL, (**c**) 1×10^{-3} g/mL, (**d**) 1×10^{-4} g/mL). The accompanying image (top) is an optical photograph of the debonded Si surface after the intentional debonding of the bonded sample of (**a**) under an ultraviolet lamp.

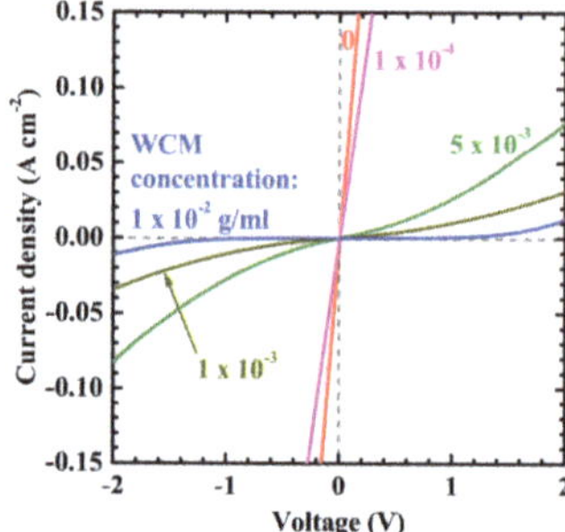

Figure 5. Interfacial current–voltage (*I*–*V*) characteristics on WCM concentration.

The inset of Figure 6 shows an optical photograph of the bonded glass plate sample, prepared for optical transmission measurements, under an ultraviolet lamp. In the snapshot of the inset of Figure 6, it is distinctly observed that the WCM at the bonded interface between the glass plates converts the incident ultraviolet light into visible blue light output. The transmittance spectrum of the bonded glass samples with WCM, along with the spectrum of the incident light, is shown in Figure 6. Note that, in our fiber-coupled optical transmission measurement setup, the sample-transmitted light scattered by the WCM particles is poorly coupled to the detection spectrometer, and therefore, the integration times of the spectrometer were adjusted so as to enable a reasonable comparison between the incident and the transmitted light. The change in the spectrum of the photon count in Figure 6 verifies that the WCM embedded in the bonded interface converted the incident light peaking at approximately 380 nm into another bundle of light peaking at approximately 460 nm. Such an optical wavelength conversion can be utilized, for instance, in solar cell applications, where the interfacial WCM absorbs an ultraviolet light (380 nm) and emits a visible light (460 nm) to fit the solar spectral irradiance to the spectral sensitivity of crystalline silicon absorption. This particular WCM can thus be utilized, for example, at the interfaces of (Al)(In)GaN/Si-based multijunction solar cells for the photocurrent improvement of the Si subcells. However, our WCM-bonding scheme can be used regardless of the material species, and thus, the most suitable combination of semiconductors can be selected, on demand, for each application situation. For example, stacked hybrid optical transmitters [4,7] in photonic integrated circuits could benefit from down-converting WCMs for longer wavelengths by enabling the use of GaAs-based or InP-based high-power diode lasers emitting at the red to the 900 nm region to be converted into the optimal wavelengths for the Si-based optical waveguides or silica-based optical fibers (1.3–1.6 μm).

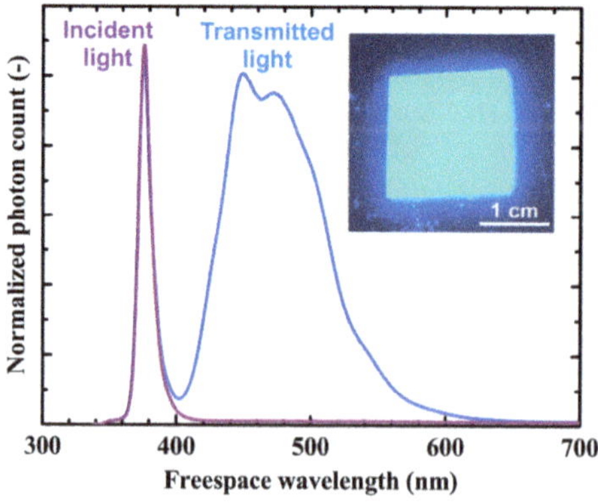

Figure 6. Transmission spectrum of the bonded samples with WCM and spectrum of the incident light. Inset: optical photograph of the bonded sample with WCM under an ultraviolet lamp.

4. Conclusions

In this study, we proposed and experimentally validated WCM-mediated semiconductor wafer bonding. This scheme could simultaneously support bond formation and interfacial function generation and thus potentially realize low-cost and high-throughput device production. The wavelength-converting heterointerfaces could improve, for example, multijunction solar cell efficiencies via photon management and current matching among the subcells and hybrid photonic-integrated circuitry performances by delivering the most suitable frequencies to each optical transceiver component. The bonding was carried out in ambient air at room temperature and therefore provides a cost advantage with regard to device manufacturing. We used Si wafers as the representative semiconductor material, for the sake of simplicity, but our concept can be extended to other semiconductors, as numerous wafer bonding experimental demonstrations between dissimilar semiconductor materials have been reported thus far [4–11]. We demonstrated the use of a down-conversion material in this study, but our bonding scheme can also be applied to up-conversion materials based, for instance, on harmonic generation [23,24] or triplet–triplet annihilation [25,26], for the efficient up-conversion of optical irradiance with a relatively low intensity, such as sunlight.

Author Contributions: K.T. conceived of the idea of the study and designed the experiments. K.T. set up the wafer bonding facility and carried out the initial bonding tests. K.K. and S.H. carried out the wafer bonding. K.K. fabricated the samples and carried out the measurements. R.I. supported the wafer bonding and the mechanical and electrical measurements. T.Y. supported the infrared imaging. K.K. and K.T. analyzed the data. K.T. supervised the study. K.K. and K.T. composed the manuscript. All the authors contributed to the discussion of the results.

Funding: This research was funded by the Japan Society for the Promotion of Science (JSPS) and the Nanotechnology Platform Project sponsored by the Ministry of Education, Culture, Sports, Science and Technology (MEXT), Japan.

Conflicts of Interest: The authors declare no conflicts of interest.

References

1. Nakamura, T. Method of Making a Semiconductor Device. U.S. Patent 3,239,908, 15 March 1966.
2. Lasky, J.B. Wafer bonding for silicon-on-insulator technologies. *Appl. Phys. Lett.* **1986**, *48*, 78–80. [CrossRef]
3. Shimbo, M.; Furukawa, K.; Fukuda, K.; Tanzawa, K. Silicon-to-silicon direct bonding method. *J. Appl. Phys.* **1986**, *60*, 2987–2989. [CrossRef]
4. Van Campenhout, J.; Rojo-Romeo, P.; Regreny, P.; Seassal, C.; Van Thourhout, D.; Verstuyft, S.; Di Cioccio, L.; Fedeli, J.M.; Lagahe, C.; Baets, R. Electrically pumped InP-based microdisk lasers integrated with a nanophotonic silicon-on-insulator waveguide circuit. *Opt. Express* **2007**, *15*, 6744–6749. [CrossRef] [PubMed]
5. Tanabe, K.; Nomura, M.; Guimard, D.; Iwamoto, S.; Arakawa, Y. Room temperature continuous wave operation of InAs/GaAs quantum dot photonic crystal nanocavity laser on silicon substrate. *Opt. Express* **2009**, *17*, 7036–7042. [CrossRef] [PubMed]
6. Tanabe, K.; Watanabe, K.; Arakawa, Y. III-V/Si hybrid photonic devices by direct fusion bonding. *Sci. Rep.* **2012**, *2*, 349. [CrossRef]
7. Crosnier, G.; Sanchez, D.; Bouchoule, S.; Monnier, P.; Beaudoin, G.; Sagnes, I.; Raj, R.; Raineri, F. Hybrid indium phosphide-on-silicon nanolaser diode. *Nat. Photonics* **2017**, *11*, 297–300. [CrossRef]
8. Wong, W.S.; Sands, T.; Cheung, N.W.; Kneissl, M.; Bour, D.P.; Mei, P.; Romano, L.T.; Johnson, N.M. $In_xGa_{1-x}N$ light emitting diodes on Si substrates fabricated by Pd-In metal bonding and laser lift-off. *Appl. Phys. Lett.* **2000**, *77*, 2822–2824. [CrossRef]
9. Park, H.; Fang, A.W.; Jones, R.; Cohen, O.; Raday, O.; Sysak, M.N.; Paniccia, M.J.; Bowers, J.E. A hybrid AlGaInAs-silicon evanescent waveguide photodetector. *Opt. Express* **2007**, *15*, 6044–6052. [CrossRef]
10. Tanabe, K.; Fontcuberta i Morral, A.; Atwater, H.A.; Aiken, D.J.; Wanlass, M.W. Direct-bonded GaAs/InGaAs solar cell. *Appl. Phys. Lett.* **2006**, *89*, 102106. [CrossRef]
11. Dimroth, F.; Grave, M.; Beutel, P.; Fiedeler, U.; Karcher, C.; Tibbits, T.N.D.; Oliva, E.; Siefer, G.; Schachtner, M.; Wekkeli, A.; et al. Wafer bonded four-junction GaInP/GaAs//GaInAsP/GaInAs concentrator solar cells with 44.7% efficiency. *Prog. Photovolt.* **2014**, *22*, 277–282. [CrossRef]
12. Naito, T.; Tanabe, K. Fabrication of Si/graphene/Si Double Heterostructures by Semiconductor Wafer Bonding towards Future Applications in Optoelectronics. *Nanomaterials* **2018**, *8*, 1048. [CrossRef] [PubMed]

13. Gibart, P.; Auzel, F.; Guillaume, J.; Zahraman, K. Below band-gap IR response of substrate-free GaAs solar cells using two-photon up-conversion. *Jpn. J. Appl. Phys.* **1996**, *35*, 4401–4402. [CrossRef]
14. Fukuda, T.; Kato, S.; Kin, E.; Okaniwa, K.; Morikawa, H.; Honda, Z.; Kamata, N. Wavelength conversion film with glass coated Eu chelate for enhanced silicon-photovoltaic cell performance. *Opt. Mater.* **2009**, *32*, 22–25. [CrossRef]
15. Hill, M.T.; Dorren, H.J.S.; de Vries, T.; Leijtens, X.J.M.; den Besten, J.H.; Smalbrugge, B.; Oei, Y.S.; Binsma, H.; Khoe, G.D.; Smit, M.K. A fast low-power optical memory based on coupled micro-ring lasers. *Nature* **2004**, *432*, 206–209. [CrossRef]
16. Urino, Y.; Shimizu, T.; Okano, M.; Hatori, N.; Ishizaka, M.; Yamamoto, T.; Baba, T.; Akagawa, T.; Akiyama, S.; Usuki, T.; et al. First demonstration of high density optical interconnects integrated with lasers, optical modulators, and photodetectors on single silicon substrate. *Opt. Express* **2011**, *19*, B159–B165. [CrossRef]
17. Matsuda, N.; Le Jeannic, H.; Fukuda, H.; Tsuchizawa, T.; Munro, W.J.; Shimizu, K.; Yamada, K.; Tokura, Y.; Takesue, H. A monolithically integrated polarization entangled photon pair source on a silicon chip. *Sci. Rep.* **2012**, *2*, 817. [CrossRef]
18. Sun, C.; Wade, M.T.; Lee, Y.; Orcutt, J.S.; Alloatti, L.; Georgas, M.S.; Waterman, A.S.; Shainline, J.M.; Avizienis, R.R.; Lin, S.; et al. Single-chip microprocessor that communicates directly using light. *Nature* **2015**, *528*, 534–538. [CrossRef]
19. Kawamitsu, S.; Nakanishi, S.; Kurogi, M.; Onouchi, H. Fluorescent Dye Compound Having Benzotriazole Structure and Wavelength-Converting Encapsulant Composition Using Same. U.S. Patent US2017210903, 27 July 2017.
20. Nam, H.G.; Nam, M.G.; Yoo, P.J.; Kim, J. Hydrogen bonding-based strongly adhesive coacervate hydrogels synthesized using poly(*N*-vinylpyrrolidone) and tannic acid. *Soft Matter* **2019**, *15*, 785–791. [CrossRef]
21. Zhang, Q.; Liu, X.; Duan, L.; Gao, G. Ultra-stretchable wearable strain sensors based on skin-inspired adhesive, tough and conductive hydrogels. *Chem. Eng. J.* **2019**, *365*, 10–19. [CrossRef]
22. Mengistie, D.A.; Ibrahem, M.A.; Wang, P.C.; Chu, C.W. Highly conductive PEDOT: PSS treated with formic acid for ITO-free polymer solar cells. *ACS Appl. Mater. Interfaces* **2014**, *6*, 2292–2299. [CrossRef]
23. Miller, G.D.; Batchko, R.G.; Tulloch, W.M.; Weise, D.R.; Fejer, M.M.; Byer, R.L. 42%-efficient single-pass cw second-harmonic generation in periodically poled lithium niobate. *Opt. Lett.* **1997**, *22*, 1834–1836. [CrossRef] [PubMed]
24. Carmon, T.; Vahala, K.J. Visible continuous emission from a silica microphotonic device by third-harmonic generation. *Nat. Phys.* **2007**, *3*, 430–435. [CrossRef]
25. Murakami, Y.; Himuro, Y.; Ito, T.; Morita, R.; Niimi, K.; Kiyoyanagi, N. Transparent and nonflammable ionogel photon upconverters and their solute transport properties. *J. Phys. Chem. B* **2016**, *120*, 748–755. [CrossRef] [PubMed]
26. Ishii, A.; Hasegawa, M. Solar-pumping upconversion of interfacial coordination nanoparticles. *Sci. Rep.* **2017**, *7*, 41446. [CrossRef] [PubMed]

MDPI
St. Alban-Anlage 66
4052 Basel
Switzerland
Tel. +41 61 683 77 34
Fax +41 61 302 89 18
www.mdpi.com

Nanomaterials Editorial Office
E-mail: nanomaterials@mdpi.com
www.mdpi.com/journal/nanomaterials

www.ingramcontent.com/pod-product-compliance
Lightning Source LLC
LaVergne TN
LVHW070618170726
843515LV00004B/119

* 9 7 8 3 0 3 6 5 2 8 6 5 6 *